CITIZENSHIP
1928

How Democracy killed the War Department Training Manual, TM 2000-25

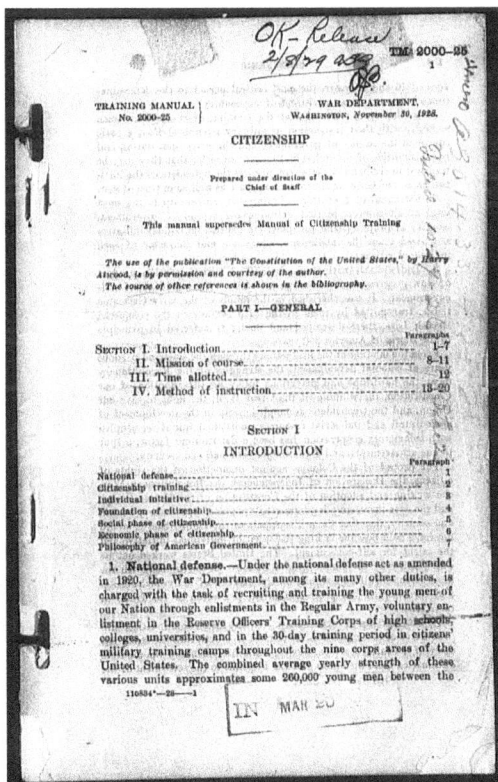

TM 2000-25, Copyedited
by James L. Tippins

RocketRanch

ROCKLEDGE, FLORIDA

I dedicate this book to my beautiful wife, Carrie, and my wonderful children, JJ and Rebecca. I truly appreciate their ongoing patience with my various interests that often lead me down many rabbit holes.

I also want to remember Harry Fuller Atwood, a great American constitutional scholar and patriot.

January 1, 1870 – December 13, 1930

"For if Men are to be precluded from offering their Sentiments on a matter, which may involve the most serious and alarming consequences, that can invite the consideration of Mankind, reason is of no use to us; the freedom of Speech may be taken away, and, dumb and silent we may be led, like sheep, to the Slaughter."

George Washington

ADDRESS TO THE OFFICERS OF THE ARMY
SATURDAY, MARCH 15, 1783

Contents

Contents
(continued)

CHAPTER ONE

Finding A 90-Year-Old American Treasure

O NE DAY, I CAME ACROSS a website posting the contents of TM 2000-25, a U.S. War Department technical manual. It explained many things about how our government works or should work. This training manual was entitled *Citizenship*, and the copy was dated November 30, 1928. The website said the main contributor was killed in 1930 by assassination during a banquet held in his honor. It also said the President ordered the manual recalled and destroyed in the early 1930s. Strong words, but was any of this true? Maybe someone made it up?

If it truly existed, the concepts within TM 2000-25 illustrate the differences in the political forms of our American government. The manual outlines the dangers of moving away from our Constitutional Republic and provides many examples from our rich American history.

Finding the documentation supporting this manual's existence took me a long time. I haven't included everything in this book, but I have verified the true history of the TM 2000-25. The Government Printing Office produced it in early 1929, and the War Department distributed it to soldiers for four years. It was also used to train civilians at various camps. And yes, it was withdrawn in 1932.

In summary, it did exist; the Government Printing Office produced over 5,000 copies before it was withdrawn from use. The manual TM 2000-25, Citizenship, was real; it was not a myth.

Who, what, and why?

The main author of the manual was Harry F. Atwood, a constitutional historian active in the early 20th century. He lived in Chicago and traveled across the United States, giving lectures at various organizations about the Constitution. His primary goal was to educate Americans about the Republic, which he called the "golden form" of government. In doing so, he informed many citizens about the nation and the leadership they were fortunate to have.

TM 2000-25 was not the first citizenship manual issued by the War Department. A citizenship manual published in 1925 did not meet the commanders' training needs in the field. In 1928, the War Department approved a new manual to replace the previous one in training classes. This manual was finished in late 1928 but was not officially released until spring 1929.

The War Department recognized that this improved manual would help soldiers understand America's true history, its early formation, and how the federal government functions. Teaching soldiers about citizenship would ensure a better citizen returning to the workforce. Non-military Americans enrolled in civilian military training programs would also benefit from these lessons.

Unfortunately, TM 2000-25 challenged the Democrats' view of democracy, which led to the manual's eventual recall. Strong democratic forces in Congress successfully worked to remove it from the War Department's training program.

Finding a microfiche copy

Finding an original copy of the manual wasn't easy, especially since it was supposedly recalled and destroyed in the 1930s. My contact with the U.S. Army Training Command yielded no results. I also searched auction websites for several years without success. However, a persistent search for "2000-25" finally paid off when I received a response one day.

A woman going through her deceased father's belongings found a box of slides labeled "TM 2000-25." She told me there were about 50 slides inside. These came from a microfiche copy of the manual stored at the National Archives in Washington, D.C. After receiving them, I digitally enlarged them to examine their contents.

I also contacted the National Archives and was told there was no record of this manual on file. The image copy I purchased looked original; I later discovered it was, in fact, a copy of the draft manual with top-level comments and approval signatures. In 2019, I found the proof copy buried inside a folder at the National Archives. It had been misfiled, and clerks couldn't locate it when I asked earlier.

My copyediting of this manual

I spent many hours copying slides from the original manual (now in the public domain) and editing them into a modern format. In some places, I kept the 1928 spelling intact. In all areas, I did not "Politically Correct" the language of that time. There may be some colorful words remaining that could offend some people today. And many Democrats probably will not like what they read.

Many PDF copies of this manual can be found online. Most lack the detail present in the original manual. Many online comments only emphasize a few paragraphs that angered the Democratic Party in the early 1930s.

This copy makes it easier for you to read the original document. My six-by-nine book format with larger fonts helps you avoid squinting at the small font in the original manual. Young soldiers usually have very good vision, and the small font in the original manual also made the smaller book easier to carry in a backpack.

My editing of Harry Fuller Atwood's work will help you explore American history in rich detail from a 1928 perspective. I hope you enjoy reading it.

What about the death of Harry Atwood?

The back cover of this book offers a teaser about the entire story. Harry Atwood wasn't assassinated; he died suddenly from a cerebral hemorrhage in Chicago on December 13, 1930. At least, that's what the newspapers report. As far as anyone knows, the President did not publicly recall the manual or order its destruction. However, the War Department had it withdrawn and stopped using it as a training aid on September 2, 1932.

Harry Atwood began writing TM 2000-25 in early 1928 after being invited to join the project by U.S. Army Chaplain C. F. Fuchter. His extensive constitutional expertise helped the War Department create an excellent training booklet. Support images for all correspondence related to this book are available on my companion website.

Oddities within the manual

The original manual uses 1920s slang. There are many points where Harry Atwood painfully describes what life was like in 1928. The American History reference he used was published around 1890. Many enemies of the Republic then are considered friends today, while many remain enemies.

A person's skin color isn't important to Atwood, and ethnicity isn't mentioned. Atwood supported the rights of all men, regardless of skin tone, and mainly focused on racial differences. He believed the main reason Black slaves remained in bondage for so long was because of American industry's business practices. In the 19th century, many Americans viewed the American Indian as the biggest obstacle to the country's full colonization.

These issues divert attention from the core topics of citizenship and American history. It's an unfortunate reality of those times. I want to say no one uses the word "Indian" to describe Native Americans anymore, but that wouldn't be true. Even the U.S. government today uses "Indian" in legal contexts.

The main points to understand from this document include those about the American colonists, the unique history of the federal government, and how citizens can help keep the United States the best place to live in the world. It might have been a while since you read a good history of our country, but I believe you'll find the American story told here truly fascinating.

On September 17, 1787, the Founding Fathers signed the U.S. Constitution. For more than 200 years, the Constitution has been the nation's highest law. President Donald J. Trump declared September 17, 2020, as Constitution Day and Citizenship Day. He also designated September 17 through September 23, 2020, as Constitution Week. This book can help you better understand our history, support our daily responsibilities as citizens, and prepare for a brighter future.

Lastly

To my knowledge, the only known original copy of TM 2000-25 available to the public is held at the Library of Congress. You may attempt to locate and replicate an original, but you are not allowed to use my text in any way except reading it from this book. If you want to use any material, please contact me for permission. Thank you for understanding.

Email: jimtippins@tm2000-25.com

Formation photograph of the American flag.
Photo copyrighted 1917.
In the Public Domain.
www.loc.gov/item/2003655430/

CHAPTER TWO

War Department Manual TM 2000-25: The Problem

H ARRY ATWOOD, a constitutional historian of the early 20th century, led the primary compilation of TM 2000-25. His previously published works contained arguments about the form of government our Nation should and should not have. This manual, written over ninety years ago, is impressive and relates to many of today's social and governmental issues.

The political problems in the early 1920s were new to America. As you read the text, remember that Harry Atwood focused on exposing the issues that might allow our great Nation to perish. He described the actions degrading our Republic and the solutions needed to keep it healthy and vibrant.

By early April 1929, approximately 7,500 copies were printed by the Government Printing Office and distributed by the War Department. Because of its "controversial nature," as described by outraged Democrats in Congress in the early 1930s, it was pulled from circulation on September 2, 1932. What led to this removal, and who was responsible for it?

To answer this, first, consider several factors. Congress had decided to limit funding of the War Department through two-year appropriations. If the Army didn't cooperate with Congressional demands, Congress would freeze the funds necessary to operate the military. We had just come out of the Great World War, and some congressmen feared the military might become so powerful it could stage a coup. But was this the main issue?

Secondly, consider politics. TM 2000-25's main issue was its definition of democracy. Today, almost everyone debates whether our government is a democracy or some variation of it. Others claim it's a constitutional representative republic. There are many diluted definitions as well.

Researching the use of "Democracy," including when it first appeared frequently in newspapers, reveals an interesting piece of data. The term was initially used mainly by Democratic Party organizations in the late 1800s to describe their platform. Give them credit for a clever political move. Say "Democracy!" every day, and the Democratic Party would stay at the top of politics. An added benefit was that people would associate the Democratic Party with democracy itself. Many would then prefer the Democratic Party to represent our American "Democracy." Again, it was a clever, if somewhat misleading, strategy.

Today, most Americans associate "Democracy" with America and the federal government. Although this isn't accurate, it has been promoted by Democrats, dictionaries, and newspapers since the late 19th century. Unfortunately, even many Republicans still mistakenly refer to our Republic as a democracy.

I found a prime example of democracy in the Bucyrus Journal, published in Bucyrus, Ohio, on Friday, October 10, 1890. This newspaper report includes statements from a speech by Daniel J. Ryan, a candidate for Secretary of State in Ohio. The article is lengthy, like most speeches by politicians. Here are a few excerpts:

Democrats would begin to use "Democracy" to symbolize an ideal Democratic federal government. Dictionaries would support this usage.

CHAPTER THREE

Democracy in Dictionaries: The Deception Is Reinforced

MOST READERS do not know the definition of democracy from early American history. They also don't know the changes in dictionaries over time. Editors can be swayed differently, especially when personal politics are considered. As I reflect and report here, I am not attacking Webster's/Merriam-Webster's vast printing organization. Realize people edit these books, and people could and did change them to suit their ideology in the past. No national group guided them while building their dictionary.

Look back at how the terms democracy and republic have evolved over time. Noah Webster's American Dictionary of the English Language, 1828, defined democracy:

DEMOCRACY, n. *[Gr. Derived from people, and to possess, to govern.]* Government by the people; a form of government, in which the supreme power is lodged in the hands of the people collectively or in which the people exercise the powers of legislation. Such was the government of Athens.

Webster's 1828 dictionary continues with Democratic terms:

DEMOCRAT, n. One who adheres to a government by the people, or favors the extension of the right of suffrage to all classes of men.

DEMOCRATIC, DEMOCRATICAL, a. Popular; pertaining to democracy or government by the people; *as a democratical* form of government.

The definition of democracy explains how rule by the people, not by representatives, controls the government. Webster also defined a republic and related terms in 1828.

REPUBLIC, n. [L. public affairs.]
1. A commonwealth; a state in which the exercise of the sovereign power is lodged in representatives elected by the people. In modern usage, it differs from a democracy or democratic state, in which the people exercise the powers of sovereignty in person. Yet, the democracies of Greece are often called republics.

2. Common interest; the public. [Not in use.]

REPUBLICAN, a. Pertaining to a republic; consisting of a commonwealth; as a republican constitution or government.

2. Consonant to the principles of a republic; as republican sentiments or opinions; republican manners.

REPUBLICAN, n. One who favors or prefers a republican form of government.

REPUBLICANISM, n. A republican form or system of government.

2. Attachment to a republican form of government. *Burke.*

It was very clear what democracy and republic meant in 1828. Webster's definition of democracy guarantees that the people control the government because "people exercise the powers of legislation."

Webster's definition of a republic describes a government in *"which the exercise of the sovereign power is lodged in representatives elected by the people. In modern (1828) usage, it differs from a democracy or democratic state, in which the people exercise the powers of sovereignty in person."*

Webster's 1828 definitions of democracy (tyranny) and republic (representation) align with the Founders' ideas of the United States. They sought to remove tyranny by the masses in favor of representation by an educated few.

Jump ahead about sixty years to *A Practical Dictionary of the English Language,* based on *The Unabridged Dictionary* of Noah Webster. This dictionary was *"EDITED UNDER THE SUPERVISION OF NOAH PORTER, D.D., LL.D., President of Yale College, BY DORSEY GARDNER"* and published in 1884.

In 1884, the definition of democracy changed:

DEMOCRACY, n. A form of government in which supreme power is vested in the people, and the legislative and executive functions are exercised by the people or by persons representing them; principles held by one of the political parties of the U. S.

DEMOCRAT, n. An adherent or promoter of, etc.

DEMOCRAT, -ICAL, a. Pertinent to, or favoring, etc.; constructed upon the principle of popular government; favoring popular rights.

What? When was "or by persons representing them" added to the definition of democracy?

In 1884, the definition of a republic was unchanged.

REPUBLIC, n. A state in which the sovereign power is exercised by representatives elected by the people; a commonwealth.

REPUBLICAN, a. Pertaining to a republic; consonant with the principles of a republic.

REPUBLICAN, n. One who favors a republican form of government; in U. S., since 1856, a member of the political party opposed to the extension of slavery.

REPUBLICANISM, n. A republican form or system of government; attachment to a republican form of government.

According to Porter, democracy in 1884 has now evolved into a representative system. You can't have it both ways, but perhaps there's a reason the definition is being altered. Could it be "The Democracy"?

Revisit Webster's 1828 definition of democracy, which clearly states that under democracy, people, not their representatives, control the government.

Here it is again:

"...the supreme power is lodged in the hands of the people collectively, or in which the people exercise the powers of legislation. Such was the government of Athens."

The definition of democracy in the 1884 edition of Webster's dictionary is like small cracks in a dam starting to spread. However, the real significant changes occurred in the 1890 edition. Many of these changes still persist today.

Noah Porter also led the editing staff for the 1890 dictionary. He was president of Yale College and a strong supporter of anti-slavery causes. While some of his focus in the dictionary supported political anti-slavery ideals, he or his staff ignored the true 1828 definition of democracy and equated it with Republicanism. This dictionary merges parts of "Republic" into "Democracy."

Here are excerpts from the preface of the 1890 edition:

In the same year (1879) a more formal beginning was made in the preparation of the edition which is now completed and will be known as the Revision of 1890...The task of adjustment is often the most difficult of all, although it may show the least of the careful attention which it has cost. All these and other difficulties can only be overcome by the employment for many years of a large number of trained assistants in the office who have devoted themselves to literary research and verbal criticism, and of a corps of specialists who have made original contributions...In the important department of Etymology, the excellent work of the last edition has been supervised and readjusted to the demands of modern Philology and recast by Professor Edward S. Shelton, of Harvard University...The Revision now given to the public is the fruit of over ten years of work by a large editorial staff, in which publishers and editors have spared neither expense nor pains to produce a comprehensive, accurate, and symmetrical work.
November 1890 - NOAH PORTER

As you'll see, the 1890 definition of democracy and republic is thoroughly revised in Porter's new volume. Modern dictionaries derived from the 1890 version are only slightly different. The 1890 definition of democracy is what Harry Atwood challenged in 1928 with TM 2000-25. Unfortunately, the 1890 definition had been unchallenged in public use and, more importantly, within the educational system for over twenty-five years when

Atwood addressed it. By then, many Americans wrongly believed we were a democracy.

Had Harry Atwood presented arguments about these changes to the definition of democracy in TM 2000-25, it would have been more difficult for people to criticize or oppose them. He did not do so and died before he could refute the Democrats in Congress.

What changed in Noah Porter's version? Consider the following definitions of interest from Noah Webster's *American Dictionary of the English Language*, 1890:

DEMOCRACY, n. [Gr. Derived from the people, to be strong, to rule, strength.]

1. Government by the people, a form of government in which the supreme power is retained and directly exercised by the people.

2. Government by popular representation; a form of government in which the supreme power is retained by the people, but is indirectly exercised through a system of representation and delegated authority periodically renewed; a constitutional representative government, a republic.

3. Collectively, the people, regarded as the source of government. — *Milton.*

4. The principles and policy of the Democratic party, so called. [*U. S.*]

Amazingly, the description in line #2 above adds the word *republic* to the definition of democracy! (Had Webster's spirit somehow become aware of this, he would have rolled his physical body over in his grave! I also think the Founders would be dismayed about this, too!) *Constitutional representative government, popular representation,* and *delegated authority* are examples of a republic, not a democracy. Why was the definition of democracy modified? Who would benefit from it? For now, continue with Noah Porter's 1890 definitions:

DEMOCRAT, n.
1. One who is an adherent or advocate of democracy, or government by the people.
> "Whatever they call him, what care I,
> Aristocrat, democrat, autocrat. — Tennyson."
2. A member of the Democratic party. [U. S.]

DEMOCRATIC, a.
1. Pertaining to democracy; favoring democracy, or constructed upon the principle of government by the people.
2. Relating to a political party so called.
3. Befitting the common people, — opposed to *aristocratic*.
The Democratic party, the name of one of the chief political parties in the United States.
(I believe the above line should have been numbered 4.)

How about the definition of a republic? Look at #2 below. It's perfect for a republic!

REPUBLIC, n. [L. commonwealth; a thing, an affair + public.]
1. Commonweal. [Obsolete] — B. *Jonson.*

2. A state in which the sovereign power resides in the whole body of the people and is exercised by representatives elected by them; a commonwealth. [Compare] DEMOCRACY, 2.

Comparing the "DEMOCRACY" definition on line #2 to the republic definition on line #2 blurs the distinction between a republic and a democracy. It tries to merge the two definitions. Was this intentional?

Noah Porter expressed his personal opinion on slavery in this dictionary. Although it accurately explains everyone's rights, such commentary is not typically included in a dictionary. It is positioned near the definition of a republic. My guess is he added it there to support his statement about the 1865 Republican Party and its stance against slavery. The editors also included the following remark within the definition of republic, complete with a "pointing finger" to draw your attention.

> ☞ In some ancient states called republics the sovereign power was exercised by an hereditary aristocracy or a privileged few, constituting a government now distinctively called an *aristocracy*. In some there was a division of authority between an aristocracy and the whole body of the people except slaves. No existing republic recognizes an exclusive privilege of any class to govern, or tolerates the institution of slavery.
>
> **Republic of letters**, the collective body of literary or learned men.

REPUBLICAN, a.
1. Of or pertaining to a republic.
   "The Roman emperors were republican magistrates named by the senate. — Macaulay."
2. Consonant with the principles of a republic; as *republican* sentiments or opinions; *republican* manners.

Next, Noah talked about his party, the Republican Party. He wanted to make it clear that the Republican Party is his preferred choice. Here is his endorsement as printed in 1890.

> **Republican party.** (*U. S. Politics*) (*a*) An earlier name of the Democratic party when it was opposed to the Federal party. Thomas Jefferson was its great leader. (*b*) One of the existing great parties. It was organized in 1856 by a combination of voters from other parties for the purpose of opposing the extension of slavery, and in 1860 it elected Abraham Lincoln president.

Here are more definitions around republic:

REPUBLICAN, n.
1. One who favors or prefers a republican form of government.

2. (U. S. Politics) A member of the Republican party.

3. [A bird]

REPUBLICANISM, n.
1. A republican form or system of government; the principles or theory of republican government.

2. Attachment to, or political sympathy for, a republican form of government. — *Burke.*

3. The principles and policy of the Republican party, so called. [*U. S.*]

As you read earlier, in 1890, "The Democracy" was used in Ohio by the Democratic Party to describe their form of government. Maybe the editors were Democrats trying to embed democracy into the educational system? Or maybe they were fooled into believing the lie?

These basic definitions of republic and democracy stayed the same for many years. Harry Atwood strongly opposed calling our federal system a democracy. Reading the manual later in this book will show how much he disliked the term.

Jump ahead to definitions randomly selected from the Merriam-Webster Dictionary, 1956 edition. I will not quote them word-for-word due to copyright restrictions by the current book owner.

Noteworthy changes include the anti-slavery message being diluted, and the development of "absolute" and "pure" democracy. "Representative Democracy" has been introduced to blend Democratic principles with Republican representation.

The definitions printed in 1956 were made during the height of the "McCarthy" era. Could it be that communists had influence over these publications?

Also, in 1956, it remained important for the dictionary to remind people that the Republican Party was founded in 1854 to oppose slavery. Black American civil rights were still under suppression in 1956, and tensions would violently surface within a decade.

Today's definitions of our interest in the 2016 printed version of the Merriam-Webster Dictionary, Merriam-Webster, Incorporated, are largely the same. There are some exceptions, but most definitions have remained unchanged since 1956.

Noteworthy changes include democracy, defined as the "rule of the majority," and all forms of representation or republican extensions of democracy are absent. Democratic has become synonymous with social reform and internationalism.

These definitions clarify how words are used in American society. The educational system embeds these ideas into the minds of our youth. Over three generations have learned this definition of democracy and now believe it is our form of government in the United States. All of us have been misled.

At the conclusion of the Continental Congress in 1789, James McHenry, a Maryland delegate to the Constitutional Convention, recorded this in his diary:

A lady asked Dr. Franklin, Well Doctor, what have we got, a republic or a monarchy — A republic replied the Doctor if you can keep it.

---

18 —

+ A lady asked D.ᵣ Franklin well Doctor what have we got a republic or a monarchy — A republic replied the Doctor if you can keep it.

+ The Lady here alluded to was Mrs. Powell of Philadᵃ.

James McHenry diary, 1787 Constitutional Convention

I don't believe Benjamin Franklin ever mentioned: "a democracy — if you can keep it."

TM 2000-25 was released to praise from the War Department, some Congressmen, and ordinary citizens. What occurred? What led to this manual being withdrawn and discontinued?

CHAPTER FOUR

TM 2000-25:
The Congressional Issue

THERE WAS SPECULATION that newspaper reporters made Congress aware of the democracy definition in the citizenship manual. TM 2000-25 may have first officially entered the House via Ross Alexander Collins (April 25, 1880 – July 14, 1968), a U.S. Representative and Democrat from Mississippi.

My impression of Collins, based on his rhetoric, is that he despises the military and can influence the War Department's financial situation.

The following excerpts are from the Congressional Record, Volume LXXII-PART 2, dated January 10, 1930. Representative Collins is on a lengthy rant about military training. Internationalism, the very foundation the Communists advocate, is the main focus of his speech. However, what he truly complains about is the definition of democracy.

I find no rebuttal from Harry Atwood in the days after Collins's complaints and until Atwood's death later that year. I would guess he was unaware of the comments made against the manual in Congress. Had Atwood lived into 1931, who knows what he might have done?

Mr. COLLINS:

TEACHING MILITARY CITIZENSHIP

The publicity which these summer camps get has so much to say about teaching the boys to be good citizens and says so little about the serious business of fighting, we may well ask what they mean by good citizenship.

Now, "good citizenship" is a broad term and is apt to mean different things to different people. To the preacher, good citizenship is apt to mean a correct attitude toward God and things godly; to the lawyer, it is apt to mean obeying the law;

To the storekeeper, it may suggest paying one's bills. No one should be surprised if the soldier thinks of good citizenship in terms of enthusiastic support for the military program and the Military Establishment.

The War Department has published an official Manual on Citizenship Training (T. M. No. 2000-25) for the use of officers teaching young men in the citizens' military training camps, the Reserve Officers' Training Corps, and so forth, which bears out this suspicion very well. It sings the praises of military training, saying:

Business invariably gives preference to the young man who has had training in military leadership. Many industries provide their employees with 30 days' vacation on pay for the purpose of attendance at a summer training camp, knowing that they will return to their employment better equipped, better disciplined, and in every way much more valuable to themselves and their employers.

It takes a slap at those who do not continually boost for a bigger and still bigger Army by referring to their attitude as "destructive idealism." I quote:

The attempt to undermine the Nation from within is more serious than the threat of armed force from without.
An impractical and destructive idealism called internationalism is being propagated by certain foreign agitators and is being echoed and reechoed by many of the Nation's "intellectuals." Its efforts are to combat the spirit of patriotism, to destroy that spirit of nationalism without which no people can long endure. • • •

TEACHING MILITARY CITIZENSHIP
(CONTINUED)

I take it President Hoover and Prime Minister MacDonald had not read this, or they would not have made the dangerous internationalistic pronouncement quoted from earlier in this speech. By this standard, the words of Jesus Christ in the Sermon on the Mount, "Blessed are the peacemakers, for they shall be called children of God," sound like the rankest Bolshevism.

MILITARY MANUAL CRITICIZES DEMOCRACY

I am disturbed by a recurring note in this official manual on "citizenship," where the General Staff seems so concerned about what they call "enemies within" the country. They come dangerously near suggesting that a class war is inevitable by continually harping on the dangers of what they call "collectivist" activities. One wonders if they are trying to strike at such old American organizations as trade-unions and such. This fear is deepened by their definition of democracy, which I quote:

Democracy: A government of the masses. Authority derived through mass meeting or any other form of "direct" expression. Results in mobocracy. Attitude toward property is communistic - negating property rights. Attitude toward law is that the will of the majority shall regulate, whether it be based upon deliberation or governed by passion, prejudice, and impulse, without restraint or regard to consequences. Results in demagogism, license, agitation, discontent, anarchy.

Why should the General Staff of our Army so characterize democracy? This is a sample of citizenship that our military men are teaching our boys. Does it look toward progress or toward militarism? Will it not aggravate the very communism it is meant to check?

Collins understands what democracy is—it's a shortened form of "The Democracy" promoted by Ohio Democrats forty years ago. He aims to defend "The Democracy" from the republic and the American people.

Later, he kept criticizing the Army about its budget and manpower. I wonder if limiting the military buildup at the time eventually strengthened Japan's decision to attack Pearl Harbor. Had we been stronger, would Japan have tried to sink our navy?

Representative Collins initiated the process by urging Democrats to recognize and eliminate TM 2000-25. He did not attempt direct action to remove the manual at that time. The War Department relied on Congress for support and could not block the congressman's efforts to discredit the manual. Collins's comments alerted the War Department that he disapproved of this entirely anti-Democratic Party document.

The following years saw more complaints to remove the manual's offensive Democratic language and stop its use in training programs. By January 1930, the booklet TM 2000-25 had been distributed for nearly a year. At least 7,500 copies were known to have been printed in 1929, and there were likely many more printings. I cannot access Library of Congress information about the total number printed.

The next time I found Congressional information about TM 2000-25 was in a 1931 document. A Non-Partisan League Republican from North Dakota, Senator Lynn Frazier, submitted a newspaper petition report. It was signed by 10,000 college students trying to avoid mandatory military training. At that time, Frazier's political party was probably the first RINO party, "Republican In Name Only," as they eventually shed their mask and became a faction of the Democratic Party.

These college students cited Atwood's definition of Democracy in TM 2000-25 as their reason for challenging mandatory military training. The Congressional record does not show whether they succeeded in avoiding the training. However, their complaint and Frazier's report to

Congress drew more negative attention to TM 2000-25. It was another "nail in the coffin" for the manual and would harm efforts to keep TM 2000-25 active in the War Department's training system.

Following is the student's petition, as documented in the March 2, 1931 Congressional Record. It states:

We, the undersigned students in American colleges and universities, protest against compulsory military training in the colleges for the following reasons:

We believe that military training courses tend to teach doctrines contrary to the principles of the American Government. In this light, we cite a definition of democracy as involving 'agitation, anarchy, discontent' from Manual 2000-25, of the War Department. We object to the use of Government funds to inculcate beliefs to which our government is unalterably opposed.

We believe that military training courses seek to idealize war and inculcate a spirit of unquestioning military obedience, which is an emotional armament of war. We quote with approval Dr. Raymond B. Fosdick's statement that military drill 'has as its chief result...a change in the mental outlook of young people so that they look upon war as a normal part of life and expect to take part in it. It habituates the thought of the participants to slaughter as a rational means of settling international difficulties as a legitimate means of reaching decisions.' We consider that our military drill courses are not only inconsistent with the Kellogg pact, repudiating war as a means of settling international disputes, but constitute a grave danger to world peace.

Ten years later, the Japanese would attack Pearl Harbor. Military training doesn't make you dull to war; it sharpens your awareness of the dangers involved. Naturally, peacetime would be ideal if it involved avoiding war or not participating in any other nation's conflicts. The best way to prevent war is to be a strong, militarily capable nation that others wouldn't want to challenge. Many modern deterrence strategies include peace through strength and mutually assured destruction. These methods have helped maintain world peace but have done little to prevent distant skirmishes.

I could find no other mention of TM 2000-25 in Congress before its recall in 1932. Maybe Congress used secret channels with the War Department to quietly suppress the manual.

TM 2000-25 was again mentioned in the Congressional Record in 1936, about four years after it was withdrawn from general circulation. Democratic members of Congress cited the manual while debating some of the War Department's 1936 technical manual content.

These statements are from the House Congressional Record VOLUME 80-PART 2, dated February 13, 1936:

Mr. MARCANTONIO. Mr. Chairman, it is my purpose at this time to point out to the membership some of the statements contained in some of the military manuals issued by the War Department. From 1928 to 1932 in Army Training Manual No. 2000-25, there appears the following official War Department definition of democracy:

Democracy: A government of the masses. Authority derived through mass meeting or any other form of direct expression. Results in mobocracy. Attitude toward property is communistic — negating property rights. Attitude toward law is that the will of the majority shall regulate, whether it be based upon deliberation or governed by passion, prejudice, and impulse, without restraint or regard for consequences. Results in demagogism, license, agitation, discontent, anarchy.

This is the definition of democracy which the War Department taught to thousands of American soldiers. If there was ever anything more subversive than this definition of democracy ever issued in any publication, I would like to know it. There is very little difference between this definition and that given to democracy by the Nazis. However, I want to state, in all fairness, that this publication was withdrawn after it had been used for 4 years, 1928-32.

I found two more references from the 1936 Congress. The following is from the Senate Congressional Record VOLUME 80-PART 4, dated March 17, 1936:

Mr. WHEELER. With reference to giving military training, it seems to me that what has happened in Japan recently ought to be a lesson to the people of the United States. The Japanese have been giving their young people military training, the result of which has been that the military element has taken over the civil functions of the government. That is one reason why, in my judgment, we should not, at this time, attempt to build up a great military force in the United States.

Mr. BONE. Mr. President, will the Senator yield?

Mr. COPELAND. I yield.

Mr. BONE. I wish to say first to the Senator from Vermont that I am not desirous of taking him off the floor. I simply want to call the attention of the Senator from New York, handling the bill, to a training manual put out by the War Department, for the use of young men in the training courses in our colleges. This manual contains some of the most extraordinary language and proposals that I have ever seen in print. This matter has been previously adverted to on the floor of the Senate by the Senator from Nebraska, Mr. NORRIS. It is not a matter of first impression in the Senate. The title of the manual is "Citizenship." It was prepared under the direction of the Chief of Staff of the United States Army, for use in the Reserve Officers' Training Camps. It bears date of November 30, 1928. I wonder why the War Department, which is merely supposed to be training officers, is telling the young men of this country things of this kind.

Senator Bone commented above, "This manual contains some of the most extraordinary language and proposals that I have ever seen in print." Of course, he would say that, but he actually means, "This manual contains some of the most damaging language to Democrats and proposals to undermine democracy that I have ever seen in print."

Senator Bone continues:

I pass to the next gem in this document. Democracy is defined, and I want my colleagues to listen carefully to this definition of "democracy" by our War Department:

A government of the masses; results in mobocracy; attitude toward property is communistic, negating property rights; attitude toward law is that the majority shall regulate.

Oh, what an astounding proposition. How un-American that the majority should by some awful mischance have anything to say about how the rules of the game are made. Quoting again:

Democracy is the direct rule of the people and has been repeatedly tried without success.

Under the representative form of government, there is no place for direct action (by the people). The inherent characteristic of a republic is government by representation. The people are permitted to do only two things; they may vote once every 4 years for the executive and once in 2 years for members of the legislative body.

One might well imagine, Mr. President, that the War Department sits in sackcloth and ashes and rends and tears its garments all the time because the people are even permitted to do that much.

Democracy—

My father spent nearly 4 years with a rifle on his shoulder, and he almost died from wounds in a military prison upholding democracy in this country. I say as a Member of the Senate that I do not like suggestions from the War Department of this country that democracy, which my father was willing to die to preserve, is a foul and noxious thing.

Senator Bone is loudly espousing the Democratic Party line. His father was willing to die for "The Democracy," not the republic. Most likely, his father had no idea democracy had overtaken the republic when he served.

Senator Copeland pointed out that TM 2000-25 had been withdrawn by order of the Secretary of War. These statements are from the official record as Congress continued on March 18, 1936:

Mr. COPELAND. The Secretary of War called attention to the fact that the Senator from Washington did not go quite far enough in his reading and did not read the final conclusion of the Department regarding the matters stated in the pamphlet on citizenship. But we do not need to consider that further, I am sure, because I have this memorandum from the War Department regarding the pamphlet from which the Senator from Washington quoted. I read:

A citizenship manual for use by instructors in the C. M. T. camps was prepared under direction of the War Department and issued for trial use in June 1927. The actual preparation of the manual was done by Chaplain C. F. Fuchter, in collaboration with the American Citizenship Foundation.

In 1928 this manual was revised by Chaplain Fuchter under the supervision of the War Department, and was distributed for use in the 1929 camps.

This is the article on citizenship referred to yesterday by the Senator from Washington:

Following its distribution, some commendatory letters were received from citizens, but there were a very large number of letters criticizing the pamphlet, which continued during the next year or two. Most of the criticism was directed toward the paragraph on democracy, which failed to be read in conjunction with the succeeding definition of a republic.

It was that matter relating to democracy upon which the Senator from Washington enlarged last night. Here is the meat of the War Department letter:

The War Department, realizing that an instructional pamphlet of such a controversial nature should not be continued in use, on September 2, 1932, directed that the manual be withdrawn from circulation and its further use as a military textbook should be discontinued.

It's time for Senator Bone to identify who was responsible for the material in TM 2000-25, as the official record continued on March 18, 1936.

Mr. BONE. I have in my hand the letter written by the General Staff of the Army, to which the Senator from New York referred a moment ago, which indicates the pamphlet to which I referred yesterday was withdrawn from the use of the camps and which indicates that the manual was prepared by Chaplain Fuchter "under the supervision of the War Department" and distributed for use in the 1929 camps. I should like to ask the Senator what is the American Citizenship Foundation, if he knows? What is that outfit?

Mr. COPELAND. I do not know.

Mr. BONE. It must be a concern which has very little use for democracy or for public ownership. I am wondering if the Senator from New York or the War Department can apprise us why they call on some outfit like that for the preparation of such a pamphlet, in view of the opinions which it entertains?

Mr. COPELAND. The only point in discussion this morning is to make clear that this pamphlet, offensive to the Senator from Washington, was withdrawn from official use in 1932, and is not now in use in any camp or organization under control of the United States Army.

Mr. BONE. I want the Senator from New York to understand it is not because it was offensive to me that I criticized it, but in my judgment, it was an assault upon the very things for which men have bled and died in this country. I cannot imagine the War Department assuming responsibility for putting out a pamphlet of that kind. What right has the War Department to attack public ownership under the guise of teaching young men how to use rifles? It is that sort of thing that makes me suspicious of military training which is made compulsory in schools.

There is no record of Harry Atwood in the Congressional Record. He was the primary contributor to the material, yet the War Department stated that Chaplain

Fuchter was responsible for the content. Since Harry Atwood died in 1930, he would not have been able to testify for the War Department before Congress. I wonder what he could have said to protect TM 2000-25 from the Democrats.

I found one more reference from 1936. There might have been more discussions of TM 2000-25 in Congress, but I have included enough to convey the main idea. Democrats disliked the definition of democracy in the manual. Without realizing it at the time, Atwood revealed their plan to keep democracy prominent in public discourse.

America fought the Spanish-American War in 1898 and the Great World War. Most citizens were wary of military build-ups after these wars. The United States participated in the Great World War for about a year and a half, losing 117,000 soldiers. Many more died from the Spanish Flu.

It's now clear that many members of Congress disliked the military. My question is, why? Were they secretly supporting some ideology other than capitalism? Were they trying to save money? Were they simply "useful idiots" in the pursuit of Communism?

Hindsight is 20/20, and the nation should have been building a strong military during peacetime. The benefit is twofold: you can respond quickly to a threat, and in most cases, the threats won't materialize because of your preparedness to fight.

There were also civilians outside Congress who disliked military preparation. The press backed them by giving them a platform to share their views. They even received publicity in the Congressional Record.

Civilians used anti-Americanism examples to urge Congress to cut some military aid. Congress ultimately restricted the War Department's funding.

The following statement is from the House Congressional Record VOLUME 80 -PART 6, dated May 7, 1936:

WE MILITARIZE
By Oswald Garrison Villard

For four years, 1928-32, the Army Training Manual No. 2000-25 carried this extraordinary definition of American democracy to hundreds of thousands of young Americans who were taking military instruction: "Democracy: A government of the masses. Authority derived through mass meeting or any other form of "direct" expression. Results in mobocracy. Attitude toward property is communistic -- negating property rights. Attitude toward law is that the will of the majority shall regulate, whether it be based upon deliberation or governed by passion, prejudice, and impulse, without restraint or regard for consequences. Results in demagogism, license, agitation, discontent, anarchy.

When this choice bit of loyalty to our American democracy and institutions was exposed to public gaze in the press, it was promptly withdrawn. When I printed it a year ago, Secretary Dern wrote me a kindly letter asking, "Why pick on a sinner after he has reformed?"

The reply, of course, is that it is an alarming state of affairs when, during four years, some officers in the War Department can put such a rank piece of disloyal, subversive anti-Americanism into a widely distributed Government handbook and that the incident must be neither overlooked nor forgotten by those who cherish their country's democratic institutions.

Five years later, on December 7, 1941, militarization would entail an effort of previously unknown scale. Over a million Americans were killed or wounded in World War II, untold thousands died indirectly, and the public's lives were changed forever.

Democracy had indeed triumphed over the republic.

CHAPTER FIVE

TM 2000-25:
Newspaper Reporting

VARIOUS NEWSPAPER CLIPPINGS reflect the disdain ordinary Democrats felt about Atwood's thoughts on democracy. Yet many editorial letters show genuine support for him and a Republican form of the federal government. Does this mean many copies of the manual may have escaped destruction in 1932? Or could there be other reasons TM 2000-25 is quoted so many times over the following 90 years?

I have gathered a small sample of excerpts from various newspapers across the United States. There are many more, but these provide a general idea of the positive and negative views of the definition of democracy in TM 2000-25. That these reports were printed also offers additional proof supporting the existence of the manuals.

The clipping dates range from 1940 to 1963. I selected various locations and editorial comments to give you a clearer picture of how the public viewed democracy and its effect on the American people.

Pensacola News Journal (Pensacola, Florida)
11 October 1940, Friday - Page 4

Republic Vs. Democracy

AN INTERESTING discussion of the difference between a republic and a democracy, terms which have become almost synonymous today, is contained in a pamphlet entitled "The March of Democracy," written by Catherine Curtis, national director of the Women Investors IN America, Inc., which describes itself as a non-profit, membership, educational organization. Its directors include many leading women of the country, but its tenor is shown by its having sponsored the Women's National Committee for Hands Off the Supreme court.

Mrs. Curtis sees the US as having strayed from the path laid out by the founding fathers, headed away from republicanism toward democracy. She says:

Today, public officials and others in this country speak of our American democracy and state that—to protect it and our liberties—we must join in the defense of the democracies. The founding fathers evidently did not intend our government to be a democracy—for that word is not found in either the Declaration of Independence or the Constitution!

U. S. Army Manual 2000-25 is no longer used in teaching our soldiers citizenship. It was withdrawn from use a few years ago, *as some of the matter contained was found to be controversial,* according to official statements!

"But we are told we entered the World War to make the world safe for democracy!"

"...democratic government has led to centralization of power, dictatorship by one man or party and *liberty* than for members of that party only. As to the United States, she says:

But we have not quite reached the final stage of democracy where "liberty" in the United States is for members of the ruling party only!"

Detroit Free Press (Detroit, Michigan)
9 September 1941, Tuesday - Page 6

Good Morning by Malcolm W. Bingay

Knocking America

Last Friday, I suggested that something was going on that had the odor of gorgonzola. I now make it stronger. Smells more like Limburger.

I contend that there was a movement in the US to bring about a form of Fascist government. Naturally, no proponent of such a plan would come out frankly and say so...the implication being a republic is something to be managed by *the better classes* while a democracy is something managed by the great unwashed mob.

MOTHERS of the U.S.A argued we are a republic, not a democracy.

ANSWER: Most certainly, the Constitution guarantees every state a republican form of government. Why not? This is a republic—a democratic republic.

So, you see, a republic can be an aristocracy as well as a democracy, but a democracy can only be one thing; a government of all the people, by all the people, for all the people.

So, when you quote the US Army Manual as saying something directly opposite, well, that is where I begin to smell Limburger.

Bingay is well-educated. It seems he has read the dictionary definition of democracy and quotes part of it. "This is a republic—a democratic republic." By now, you can see how over fifty years of misunderstanding about democracy can sour the public, political, and editorial environment.

Daily News (New York, New York)
19 April 1949, Tuesday - Page 220

"CAPITOL STUFF by JOHN O'DONNELL

The ghostwriters for President Truman in the White House—like the earlier ghostwriters of Woodrow Wilson and the departed F. D. R.—are still pounding away on the Socialist-Communists line to put across the idea that this republic of ours is a *democracy.*

By twisting and wrenching the meaning of words, they seek to put across the idea that a citizen who is "undemocratic" runs afoul of the fundamental concepts of the Declaration of Independence, the Constitution of the US, or the writings of the founding fathers—Washington, Jefferson, Monroe, Madison, Hamilton and the rest.

Reader mail has been pouring in at a surprising rate since we pointed out a few days ago that this republic is not and was never intended to be a democracy: that its founders took pains to see that it couldn't be a democracy and that the American way of life (if the founding fathers are to be believed) was frankly non-democratic and even anti-democratic.

And so, the New Englander Elbridge Gerry joined the anti-democrats of Virginia and the rest of the South when he told the assembly:

The evils we experience flow from an excess of democracy. The people do not want virtue but are the dupes of pretended patriots.

And then Hamilton moved in to put the quietus on the shouters for a pure democracy, who had praised it as the perfect government:

Experience has proven that no position has proved more false than this. The ancient democracies, in which the people themselves deliberated, never possessed one feature of good government. Their very character was tyranny; their figure deformity."

"Non-democratic and even anti-democratic." Perfect. People recognized that democracy was wrong. But just like a pest, it can be hard to eliminate.

The Courier-News (Bridgewater, New Jersey) 12 May 1956 - Page 6

When did the US become a democracy, and by what process? What happened to the Republic? This inquiry is suggested by a photostat of our soldiers' official doctrine in 1928 and what they are being taught today.

U.S. Army definition of democracy, 1928:
Democracy—A government of the masses...
U.S. Army Training Manual
No. 2000-25, 1928 p. 91

U.S. Army definition of democracy, 1952:
Meaning of democracy.
Because the US is a democracy, the majority of the people decide how our government will be organized and run—and that includes the Army, Navy, and Air Force. The people do this by electing representatives, and these men and women then carry out the wishes of the people.

The Soldier's Guide
Department of the Army
Field Manual, FM-13, June 1952

If this is a democracy, then, according to the philosophy of the Father of Philosophy, the U.S.A. is surely headed toward a fate similar to that of the historical end of ancient Rome and the Dutch Republic.

If this is a democracy, when did it happen, and what are we going to do about it?

Many people were aware of the dangers, but by 1956, the public had been educated in Democracy for nearly eighty years.

Centuries ago, democracy contributed to the downfall of the Roman people (the emperors and the Roman Senate played a small role), and some Americans worried that communism, aided by democracy, could destroy the United States.

Anderson Herald (Anderson, Indiana)
5 April 1963, Friday - Page 4:

"A Republic and A Democracy

Just to remind ourselves of what America once stood for and the type of government we had, it might be a good idea to mention here a few paragraphs that were published in Training Manual 2000-25..."

"This manual was withdrawn from the GPO, and copies were recalled shortly after the bank holiday in the early 1930s.

But after reading the definitions (of Democracy), we are now wondering just why it was withdrawn and, for that matter, what type of government we have today."

Many other newspaper clippings focus on the War Department manual TM 2000-25 in letters and editorials. But again, how did people know about the manual? It didn't exist after 1932, yet people reference it to this very day. I hope to find a hard copy someday that I can have for my very own. Until then, I have my slides, which I use to include the manual's text in this book.

CHAPTER SIX

TM 2000-25: The Demise

The Democratic powers of the purse over the War Department killed *Citizenship*. Very good training lessons were discarded even though the manual was correct about the pitfalls of democracy. A memo to the Adjutant General dated September 2, 1932, ordered the withdrawal from circulation and the discontinuance of TM 2000-25:

Subject: Training Manual 2000-25, November 30, 1928
The Secretary of War directs: -

I. That Corps Area and department Commanders be informed by letter as follows:

Training Manual 2000-25, November 30, 1928, is withdrawn from circulation and its further use as a military textbook will be discontinued.

II. That TR-2000-25(typo) will not be listed in TR 1-10 when the latter is next revised.

Edgar T. Collins,
Major General,
Assistant Chief of Staff

And with that, the manual of Citizenship, TM 2000-25, is consigned to history. I haven't found another physical copy anywhere else. The only copy I am certain exists is in the Library of Congress.

In summary, I believe the Democratic Party is responsible for ending TM 2000-25 and for the ongoing use of democracy to describe our federal government. The Democratic Party's desire to use democracy as their party platform needs to be addressed, and the term democracy should be removed from the vocabulary of those describing our Federal Government.

Democracy rears its ugly head and harms the nations that adopt it. Its venom is toxic to free people. Democracy begins a vicious cycle of chaos and tyranny. We must avoid using democracy as "anything" to describe our republic. Doing so will be a key step in correcting the wrong direction our great nation has taken since Democrats started down this path in the 1890s.

Students need to be told the truth about our political system, and adults should recognize the harmful path they've been on for more than a century. It's not too late to save the republic.

I have spent many hours reading public newspapers and scholarly books claiming we have become a Representative Democracy, a Democratic Republic, or just a direct Democracy. Those sources are incorrect, and their efforts seem to support the Democratic Party, whose political goal appears to be undermining our republic. It's important to remember that the Founders intended for us to be a republic.

As you read this manual, think of it as an introduction to how our government should operate and how its citizens should behave. The manual is written at a high-school level of understanding regarding citizenship; read it like a history book and consider its ideas carefully.

One day, I hope to carry this manual ninety years into the future to expand our knowledge of the republic and modern history. Until then, please finish this book to study the original War Department Technical Manual 2000-25. Learn and understand citizenship concepts as Harry Atwood presented them.

Together, we can make America the perfect republic the Founders envisioned. Never forget Collins's 1930 attempt to slander the manual in Congress and defend "The Democracy" and the Democratic Party. The Congressman's shameful record of destructive actions against the republic remains forever in the journal of the American Congress.

I hope this book will garner new public support for TM 2000-25. Perhaps the manual's words will reach the editors of pro-Republican papers for the first time in over ninety years. Perhaps the lie of Democracy will be exposed once again to the American public.

The truth is impossible to hide forever, or at least I like to think so.

James L. Tippins

TM 2000-25
The Original Manual

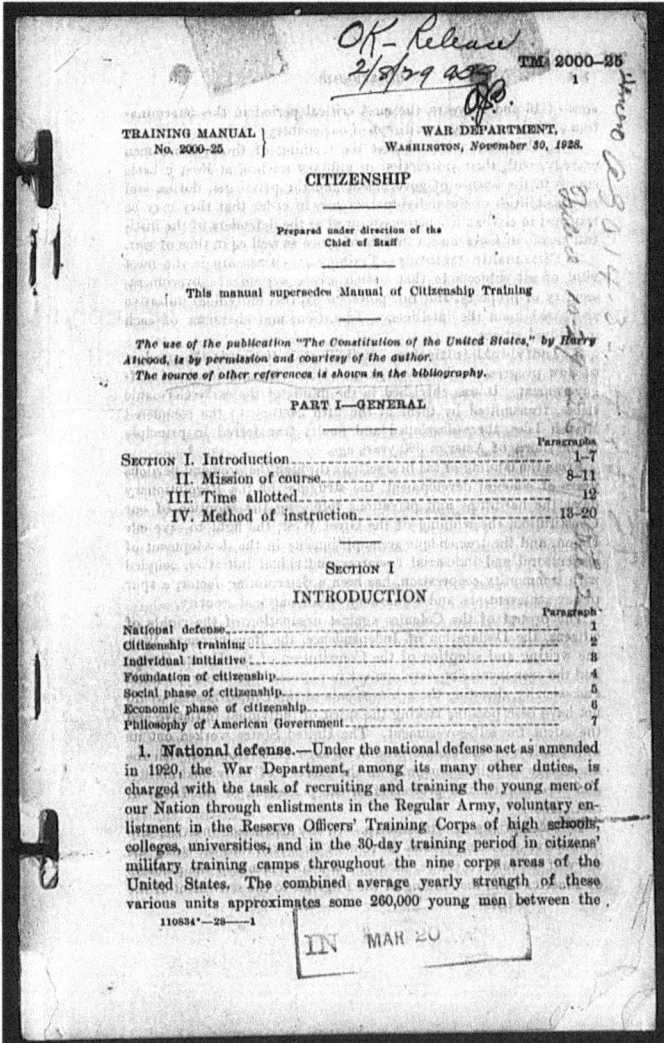

OK - Release
2/8/29

TM 2000-25
1

TRAINING MANUAL }   WAR DEPARTMENT,
No. 2000-25   }   WASHINGTON, *November 30, 1928.*

CITIZENSHIP

Prepared under direction of the
Chief of Staff

This manual supersedes Manual of Citizenship Training

The use of the publication "The Constitution of the United States," by Harry Atwood, is by permission and courtesy of the author.
The source of other references is shown in the bibliography.

PART I—GENERAL

SECTION I
INTRODUCTION

1. **National defense.**—Under the national defense act as amended in 1920, the War Department, among its many other duties, is charged with the task of recruiting and training the young men of our Nation through enlistments in the Regular Army, voluntary enlistment in the Reserve Officers' Training Corps of high schools, colleges, universities, and in the 30-day training period in citizens' military training camps throughout the nine corps areas of the United States. The combined average yearly strength of these various units approximates some 260,000 young men between the

110834°—28——1

IN   MAR 20

Original draft cover page, dated November 30, 1928.
Copyright © 2017 by James L. Tippins.

PART I – GENERAL

SECTION I — INTRODUCTION

1. **National defense.** — Under the national defense act as amended in 1920, the War Department, among its many other duties, is charged with the task of recruiting and training the young men of our Nation through enlistments in the Regular Army, voluntary enlistment in the Reserve Officers' Training Corps of high schools, colleges, universities, and in the 30-day training period in citizens' military training camps throughout the nine corps areas of the United States. The original average yearly strength of these various units approximates some 260,000 young men between the ages of 16 and 25 years, the most critical period in the determination of their real value as citizens of our country.

2. Citizenship training. — Training in citizenship is the most vital of all subjects to that nation whose system of government, security of property, and full power to express individual initiative are based upon the intelligence, education, and character of each individual citizen.

It is, therefore, essential that the training of these young men embody, with their instruction in military science, at least a basic course in the science of government and the privileges, duties, and responsibilities of the individual citizen, in order that they may be returned to civilian life better equipped as the defenders of the institutions of our Government in time of peace as well as in time of war.

3. Individual initiative. — Individual initiative is the product of slow progress in the development of the idea and ideals of self-government. It was cherished in the minds of the early Germanic tribes, transmitted by them in the fifth century to the conquered British Isles, there developed and finally transferred in principle to the shores of America 300 years ago.

From the landing of the first settlers through the slow and perilous years of colonial development, the struggles of the Revolutionary days, the hardships and privations following the adoption of our Constitution, the winning of the Great West, the fight to save our Union, and the tremendous accomplishments in the development of agricultural and industrial resources, individual initiative, coupled with community cooperation, has been a determining factor, a spur to our achievements, and a guaranty to our national security.

The protest of the Colonies against usurpation of the rights of citizens, the Declaration of Independence, the Revolutionary War, the writing and adoption of the Constitution of the United States, and the ever-increasing development in population, industry, wealth, and security, denoting the achievements of the United States, would not have been possible lacking the spirit of individual initiative and the talent for self-government. The United States worked out its own destiny by the simple process of hard labor inspired with the knowledge of full opportunity in the exercise of individual ability, and sure reward and protection in the possession of the fruits of their labor.

4. Foundation of citizenship. — In any instruction in citizenship productive of lasting results, there must be woven into the study the story of the faith, sacrifice, service, and achievements of the pioneers of America from the landing of the Pilgrims to the settlement of the Great West and the development of our vast national resources. This story, pregnant with hope, faith, courage, and the will to work, is the rock foundation upon which to build the structure of citizenship in the youth of to-day that the future may be assured in perpetuity of the institutions, principles, ideals, and traditions the development of which has made the United States great among the nations of the world.

A study of the census reports of the United States, particularly during the past 50 years, reveals a condition that to every thinking man and woman is fraught with grave danger to the continuation and maintenance of our constitutional form of government and the blessings of liberty which we enjoy. We must be prepared to recognize this situation and find the solution of the problem.

5. Social phase of citizenship. — As the result of the changing life stream of America, there has arisen one of the greatest problems of our national life. Up to 30 years ago, approximately 90 percent of all immigration to America was of Anglo-Saxon origin, that race of people which has been working out the problem of self-government for nearly 2,000 years. Due to the remarkable impetus given to industrial development following 1890, opportunity for employment was offered and every inducement made to secure the immigration of European common labor, resulting in an immediate change in the type of immigration to America, by which central, eastern, and southern Europe increased their totals by over fifty times in the 50 years from 1870 to 1920.

The history of the nations from which this later immigration originated is that of large cultural advantages in art, literature, and science, enjoyed by the ruling and favored minority, while oppression, privation, and suffering were endured by the great majority of their subjects.

This latter class, without knowledge of self-government, denied the opportunity for self-development, eagerly responding to the call of American opportunity, emigrated to our shores, here to enjoy full participation in the rights of American citizenship without a proper understanding of the meaning of liberty or the nature and value of our free institutions, the very foundation of which is laid in intelligent and active participation in government by our individual citizens.

A course of instruction in citizenship to be effective must develop the social phase of citizenship and be particularly directed to the native and foreign-born youth, setting up a clear understanding of this great problem of assimilation and amalgamation of the bloods of all nations into the virile life stream of America.

6. **Economic phase of citizenship.** — The industrial achievements of America have become the marvel of the world. Therefore, the economic phase of citizenship must be developed with careful study and with all the wisdom we possess that we may assure continued progress to the welfare, tranquility, and enrichment of our own citizens and at the same time, steer a safe course for our ship of

state in the maelstrom of world envy engendered by a knowledge of our wealth and power.

In the accomplishment of our industrial achievements, the United States has reinvested its profits in the development of horsepower, automatic machinery, labor-saving devices, transportation, communication, organization, administration, and, since World War I, has given further impetus to its accomplishments by sharing more and more the fruit of her industries with the wage-earning class. In the progress thus made, the demand for brains to replace brawn has been an ever-increasing factor in the production of our goods as to quantity and quality in order to maintain our sense of well-being, high standards of living, and to meet the competition of the world at large.

A course of instruction in citizenship must emphasize the necessity of the education of our masses as an economic measure in supplying the great need of modern industry.

7. **Philosophy of American Government.** — The philosophy of government, as set up under our Constitution, finds its keynote in individualism as opposed to that misguided philosophy of government, collectivism, which makes the State paramount in its demands over the inalienable rights of its individual citizens. Incomprehensible as it may seem, the political problems of America and of the world at large are embodied in this question of individualism as opposed to collectivism as the philosophy of government for the future development and welfare of nations.

Emphasis must be laid upon the benefits and advantages accruing to each individual citizen of our country under the form of government set up as the supreme law of the land in the Constitution of the United States of America.

SECTION II — MISSION OF COURSE

8. General purpose. — This course in citizenship is designed to teach the fundamental principles upon which our Government is founded, including an insight into the social and economic elements upon which our civilization stands. Special emphasis is laid upon the meaning of "liberty," as interpreted by the founders of this Republic, and the larger relationship of the individual citizen to others and to his Government, defining loyalty and national responsibility in terms of citizenship, recognizing that an intelligent and informed people is a greater asset than are the unintelligent, uninformed, or misinformed, and that no government can exist upon a plane higher than the moral character of its people.

9. Knowledge, the safeguard of our Republic. — Because of the rapid increase in our population, largely made up of immigrants from all parts of the world, the tendency within the family and the school is to neglect the training of our youth in the knowledge of his Government and his individual responsibility. It cannot be expected that foreign-born parents, lacking knowledge or inspiration of American ideals, will be either fitted or inspired to give such instruction to their own children.

The indifference or the neglect of native-born citizens concerning the training of their children to meet the responsibilities of citizenship is largely caused by lack of information and proper understanding of the history, ideals, and underlying principles of our political institutions.

The remarkable development of industry in America has caused a congestion of population in our large cities, creating social, economic, and political problems that materially affect the structure of our Government.

The solution of the problems of citizenship lies largely in the education of the youth of America in the principles of representative government and their personal responsibility in perpetuating and improving her free institutions.

10. Character building. — The ever increasing wants as compared to the needs of humanity, the added individual burdens and problems of modern civilization, emphasizing material rather than ethical and spiritual attainment, are tending to break down the character of our youth.

It is the mission of this course to specially emphasize the moral aspects of citizenship— to build up home discipline, reverence for religion, and respect for constituted authority.

11. National defense. — Education and training in citizenship form a vital part of national defense. It will be the mission of this course to interpret national defense through a broad and comprehensive instruction in citizenship, stressing the responsibility of the individual citizen to become fully prepared for the defense of his country in any emergency that may arise, whether of domestic or foreign import, in peace or in war.

SECTION III — TIME ALLOTTED

12.   Time allotted. — In this course of citizenship, adequate time will be allotted for instruction, arranged in a number of short periods of not more than 40 minutes duration each, which may be supplemented by addresses and travelogues illustrated with stereopticon slides, covering outstanding phases of American history, given to combined groups at such time and frequency as directed by the camp commander, with special reference to rainy day schedules.

SECTION IV — METHOD OF INSTRUCTION

13. Outlined topics. — This course will be given under a series of outlined topics briefly presented by the instructor, preceded by a few succinct historical statements bearing upon the development of our country.

14. Questionnaire. — Brief questionnaires, containing a number of questions pertinent to the subject matter contained in each lesson, are given as an aid to the instructor in guiding the general discussions by the students.

15. Subject matter suggestive. — This course is not intended to teach the details of American history, but to give special emphasis to pertinent facts and principles associated with the foundation, development, and preservation of our Government as to its social, economic, and political phases. The instructors should briefly explain the historical and psychological aspects of the various forms of government.

The subject matter and illustrations are suggestive only and are given as guides in teaching the fundamental principles of government and citizenship. The instructor will make application of these principles in such a manner as to stimulate individual thinking, leaving it to the student to reach his own conclusions based upon the facts and situations discussed.

16. Plan of instruction. — In the presentation of this course, it is necessary for the instructor to give certain definite and concise information concerning the outstanding characteristics of our country; the fundamental principles of our Government; the spirit and will to do by which it attained its present position; emphasizing the encouragement, assistance, and protection granted every individual citizen as guaranteed in our Constitution as the supreme law of the Nation; developing the idea of individual responsibility and intelligent participation in government as an economic necessity as well as an evidence of patriotism and loyalty to our country.

The didactic method concerning facts of history, social changes, economic development, and basic principles of our Government will be used without discussion and without argument, *special emphasis being given to the fact that the United States is a Republic, not a democracy.*

> **Copy Editor note:** Someone annotated this phrase in the original proof copy. Was it done during the turmoil in Congress or before publication?

Group discussions will be led by the instructor covering the cardinal points of each lesson as outlined in the text, care being exercised to confine the discussion to the limits of the lesson.

17. Selection of instructors. — There shall be designated a director of citizenship training for each Citizens' Military Training Camp. Under his supervision, company officers carefully selected by the camp commander will act as instructors in this course.

18. Suggestions for instructors. — Instructors are particularly cautioned to confine instruction and discussion in each study period not only to the lesson text but also to keep it within the scope of the general division (social, economic, political) to which that particular lesson is related. The tendency is to wander away into a discussion of all three phases of citizenship because of the close interrelationship existing in all the lessons. Clarity of instruction can be had only through close observance of this suggestion.

The instructor must use language simple enough to be readily understood by all.

The text of these lessons is so arranged as to permit additional time for study and discussion when such opportunity is available through accommodation to rainy-day schedules.

19. Supplemental instruction. — At the discretion of the camp commander, instruction may be supplemented by addresses given by selected speakers to the combined student body on subjects related to citizenship.

As a part of this course, historic facts and brief statements taken from the speeches and writings of distinguished Americans may be projected on the screen immediately preceding the feature picture at all motion-picture shows.

20. Efficiency. — To secure the most efficient results, the officers detailed as instructors should be thoroughly trained in the method of using the various studies in citizenship and the questionnaires.

A refresher or normal course will be conducted in each camp for the instruction of the designated instructors in subject matter and method of presentation, with the view of having the classes in citizenship faced by instructors as alert, competent, and as confident as are the platoons in the military drill.

PART II — COURSE OF INSTRUCTION

SOCIAL

ECONOMIC

POLITICAL

PATRIOTIC

SECTION I

LESSON 1. — THE AMERICAN CITIZEN

21. Definition of citizenship. — Citizenship is that membership in a nation which includes full civil and political rights, subject to such limitations as may be imposed by the government thereof.

22. Origin of citizenship. — Citizenship as we understand it to-day is the result of centuries of social, economic, and political experiments, in which improvement in human relations has slowly developed the idea of the benefits of governmental rules and restrictions for the protection of the rights of persons and property.

> Ancient Greece was composed of a number of city-states, each one independent of the other and conferring certain privileges upon its subjects. The greatest advantages of citizenship among these city-states was conferred by the Athenians, limited, however, to native sons of native fathers and mothers, excluding from such privileges foreigners and slaves. The Athenian ideal of citizenship was philosophical rather than practical.

It was left to the Romans, in succeeding centuries, to develop the more practical phases of citizenship, i.e., safety of the Republic, public service, stern simplicity, devotion to duty.

Above all other duties and obligations was placed that of unselfish duty to the state. It was this Roman virtue of loyalty to public duty, this devotion on the part of the citizen to the interest of the state, that, more than any other quality of the Roman character, helped to make Rome great.

Roman citizenship was confined to a privileged class, native or adopted.

In the Anglo-Saxon races, there was slowly developed the idea and ideals of self-government and of individual worth, in contrast with the earlier Greek and Roman domination of the state over the individual.

Out of these experiments in government and human relations, there has been evolved the ideals and principles of American citizenship.

23. Source of American citizenship. — The source of American citizenship is found in the Constitution and subsequent Federal enactments.

24. Acquisition of American citizenship. — American citizenship is acquired in two ways:

By birth.

By naturalization.

Birth. — For 150 years following the first settlement of the American Colonies, their inhabitants were citizens and subjects of a foreign power.

With the successful conclusion of the Revolutionary War, terminating with the treaty of peace, 1783, all persons born in the United States before the Declaration of Independence could be regarded as American citizens.

By the civil rights act of 1866, it was provided that —

All persons born in the United States and not subject to any foreign power, excluding Indians not taxed, are declared to be citizens of the United States.

By the fourteenth amendment to the Constitution —

All persons born or naturalized in the United States and subject to the jurisdiction thereof are citizens of the United States and of the State wherein they reside.

It has been decided by the Supreme Court of the United States that the children of domiciled aliens born in the United States are citizens under the fourteenth amendment. This is also true of the children of alien parents' ineligible for citizenship through naturalization.

Immigration and naturalization. — Under the Constitution, Congress is given the power over both immigration and naturalization.

In order to determine their fitness to enter the United States, each immigrant, on his arrival, is subjected to a physical and mental examination by officers of the Public Health Service.

Under the immigration act, the following classes of persons are excluded from entering the United States:

Idiots. (*Seriously, these are on the list in 1928!*)

Insane.

Epileptics.

Paupers and persons likely to become a public charge.

Professional beggars.

Persons suffering from tuberculosis or other dangerous or loathsome contagious diseases.

Persons physically or mentally so defective as to be unable to making a living.

Persons convicted of a crime or misdemeanor involving moral Turpitude.

Polygamists.

Anarchists.

Women or girls imported for immoral purposes and persons aiding in their importation.

Contract laborers — that is, those induced to migrate by offers or promise of employment or by agreement, except artists and professional men.

Children under 16 years of age unaccompanied by their parents.

With certain exceptions, no alien ineligible for citizenship is admissible to the United States.

All aliens brought into the country in violation of the law are, if possible, immediately sent back to the country whence they came on the vessel bringing them, at the expense of the vessel owners.

There is also a heavy fine upon the transportation company or vessel owner for unlawfully introducing immigrants into the United States.

Because of the great influx of non-assimilable people, which tended to lower American standards of living, and to better develop a homogenous body politic, Congress, in 1923, passed the immigration restriction act.

The abnormal immigration to America is shown in the census returns of 1900, 1910, and 1920, as follows:

1900 -- 3,687,564

1910 -- 8,795,386

1920 -- 5,735,811

The law governing immigration provides that the annual quota from each country until July 1, 1927, is 2 percent of the number of foreign-born persons of such nationality resident in continental United States as shown by the 1890 census, but the minimum quota of any nationality shall be 100.

The quota for each fiscal year thereafter will be based on a total immigration of 150,000.

The annual quota of any nationality for the fiscal year beginning July 1, 1927, and for each fiscal year thereafter, shall be a number which bears the same ratio to 150,000 as the number of inhabitants in continental United States in 1920 having that national origin (ascertained as hereinafter provided in this section) bears to the number of inhabitants in continental United States in 1920, but the minimum quota of any nationality shall be 100. — *Immigration laws, 1927.*

Under the Articles of Confederation, the power of naturalization was in the States, thereby creating confusion through the lack of uniformity in conferring citizenship.

The authority for naturalization is to be found in the Constitution and Federal laws.

The Constitution has accordingly, with great propriety, authorized the General Government to establish a uniform rule of naturalization throughout the United States. — *Madison. Constitution, Article I, section 8, paragraph 4, fourteenth amendment. Naturalization Laws.*

Under the Constitution, two methods of naturalization have grown up:

(1) By the general act of Congress conferring citizenship upon a whole class of persons, such as tribes of Indians, and the inhabitants of a new territory, like Hawaii, acquired by the United States.

(2) The general and more usual method is prescribed by the Revised Statutes, which requires the fulfillment of certain conditions before final admission into citizenship.

R. S. 381. *Oath renouncing foreign allegiance and to support constitution and laws.* — He shall, before he is admitted to citizenship, declare on oath in open court that he will support the Constitution of the United States, and that he absolutely and entirely renounces and abjures all allegiance and fidelity to any foreign prince, potentate, State, or sovereignty, and particularly by name to the prince, potentate, State, or sovereignty of which he was before a citizen or subject; that he will support and defend the Constitution and laws of the United States against all enemies, foreign and domestic, and bear true faith and allegiance to the same. — *June 29, 1906, Ch. 3592, sec. 4, 34 Stat. 596.*

R. S. 382. *Evidence of residence, character, and attachments to principles of Constitution; evidence of witnesses.* — It shall be made to appear to the satisfaction of the court admitting any alien to citizenship that immediately preceding the date of his application he has resided continuously within the United States, five years at least, and within the State or Territory where such court is at the time held one year at least, and that during that time he has behaved as a man of good moral character, attached to the principles of the Constitution of the United States, and well disposed to the good order and happiness of the same. In addition to the oath of the applicant, the testimony of at least two witnesses, citizens of the United States, as to the facts of residence, moral character, and attachment to the principles of the Constitution shall be required, and the name, place of residence, and occupation of each witness shall be set forth in the record. — *June 29, 1906, Ch. 3592, sec. 4, 34 Stat. 596.*

25. No dual allegiance. — Every alien should become a citizen in order that he may vote and hold office, and in all ways take an active part in developing, building, and maintaining the Government—national and local—that protects him.

There can be no divided allegiance here. Any man who says he is an American, but something else also, isn't an American at all. We have room for but one flag, the American flag, and this excludes the red flag, which symbolizes all wars against liberty and civilization, just as much as it excludes any foreign flag of a nation to which we are hostile.

We have room for one soul, loyalty, and that is loyalty to the American people. — *Roosevelt.*

26. Dual citizenship. — The Supreme Court declares that there are two kinds of citizenship, State and National.

Citizens of the United States residing in any State enjoy the rights of both State and United States citizenship.

In the protection thereof, we look to the National Government if the source of such rights lies in the Constitution and laws of the United States; and to the State government if such rights are based upon the constitution and laws of the State.

Dual citizenship does not imply a divided allegiance. While a State commands allegiance of its citizens, the paramount allegiance is to the Union.

> Liberty and Union, now and forever, one and inseparable.
> — *Webster.*

27. Right of suffrage. — Under the Constitution, the National Government confers American citizenship, but it is left to the States to determine who may vote at both its own and national elections.

> Constitution, Article I, section 8, paragraph 4; fourteenth and fifteenth amendments.

In America, public opinion is the ultimate force of Government. It is the expression of the mind and conscience of the whole Nation, without respect to sectional or partisan alliances.

Under the Constitution, voting is the only means provided for the expression of public opinion—it is the exercise of the will of the citizen in the protection of his rights.

28. Guaranties as to person and property. — The United States is composed of 48 sovereign States, each State having its individual constitution and laws. Yet no State may discriminate against the rights and privileges of the citizen of any other State as to person or property.

Among these guaranties are —

Opportunity for education and individual improvement.

Unrestricted possession of property.

Joint rights to interstate commerce, communication, and transportation.

Public utilities.

Freedom of residence and choice of occupation.

Care or protection on the high seas or abroad through passport privileges and international law.

29. Obligations of citizenship. — Active citizenship is gained only by becoming an enfranchised citizen of a State. This carries with it the obligation of a clear understanding of the principles of government and the courage to demand that these principles be not abridged.

Andrew Jackson said that every good citizen makes his country's honor his own, and not only cherishes it as precious, but sacred.

Lincoln declared: "I must stand by anybody that stands right; stand with him while he is right; and part with him when he is wrong."

It is essential that the individual citizen —

Exercise his right of franchise — vote — as his Paramount duty at all elections.

Uphold the Constitution as the one assurance of the security and perpetuation of the free institutions of America.

Practice self-government to assure good government for all.

Respect the rights of others, to assure the enjoyment of his own.

Contribute to the maintenance of his Government by the payment of taxes.

Obey the law as the first essential to law enforcement.

Place service to country above service to self.

Conform his conduct to the best interests of society.

The opportunities and privileges of the American citizen are limited only by his individual ability, his personal habits, and conformity to necessary legal regulations.

It is your obligation to exercise —

Care in your choice of occupation.

Diligence in preparation for your task.

Thrift to ensure advancement and prosperity.

Judgment in selection of companions.

Integrity, honor, initiative, self-reliance, self-control.

30. I am an American. ⊢ "I am an American" is a challenge to the highest ideals and aspirations of mankind; to self-sacrifice and devotion; to loyalty and patriotism; to joyful work and courageous achievement; to magnanimity and charity to all and malice to none; as we seek to uphold and perpetuate the principles of our great Republic.

I live an American; I shall die an American, and I intend to perform the duties incumbent upon me in that character to the end of my career. I mean to do this with absolute disregard of personal consequences. What are the personal consequences? What is the individual man, with all the good or evil which may betide him, in comparison with the good or evil which may befall a great country, and in the midst of great transactions which concern that country's fate? Let the consequences be what they will. I am careless. No man can suffer too much, no man falls too soon, if he suffers, or if he falls in the defense of the liberties and Constitution of his country. — *Daniel Webster.*

James L. Tippins

In the days of the Caesars, "I am a Roman citizen" was a proud exultant declaration. It was protection. It was more — it was honor and glory. Twenty centuries of advancing civilization have given to the declaration "I am an American," a higher and nobler place. It stands to-day in the forefront of earthly titles. It proclaims a sharing in the greatest opportunities. It is a trumpet call to the highest fidelity. It is the diploma of the world, the highest which humanity has to bestow. — *Judge Brewer of the Supreme Court.*

QUESTIONNAIRE — THE AMERICAN CITIZEN

Define "citizenship."
Describe the development of the idea of "citizenship."
What is the source of "American citizenship"?
How is "American citizenship" acquired?
What is the status of the children of domiciled aliens born in the United States?
Who has power over immigration and naturalization?
To what examination is the immigrant subjected on his arrival?
What classes of persons are excluded from the United States by the Immigration Act?
What disposition is made of immigrants belonging to the restricted classes?
To whom is the execution of the Immigration Laws entrusted?
What was the significance of the immigration to America by the census returns of 1900, 1910, and 1920?
What has Congress done to limit immigration? Why?
What is the source of the authority for naturalization?
Explain the provision for naturalization under the Articles of Confederation. Under the Constitution.
What is the attitude of the United States toward "dual allegiance"?
Explain the meaning of "dual citizenship."
What is the function of "public opinion"?
Who has power over the right of suffrage?
What guaranties, as to person and property are provided
the citizen by the Federal Government?
Name several obligations of citizenship.
Why ought an alien become a citizen?
Why should every citizen vote?

SECTION II

LESSON 2. — INTERDEPENDENT RELATIONSHIPS

31. Development of civilization. — Civilization had its beginning in the establishment of the family, then in the grouping of families, tribes, states, and nations.

Through these various stages, there was developed a crude order of society based primarily upon the will of an outstanding individual with power to enforce that will by control of physical forces and the means of livelihood. Thus, was established the basis of society, imperfect in its form, inadequate in its results, yet containing the essential elements for refinement and progress, viz, social interaction, protection, and advantages.

32. Mutual relationships. — In the beginning, lacking means of communication and transportation and confining efforts principally to the production of mere necessities of life, individuals and groups lived largely independently of each other.

With increasing wants, the result of enlightened intellect, with increasing facilities in transportation and communication, with development of ability for invention and improvement, independence gave way to interdependence to such a degree that today the welfare of every individual is woven into the fabric of modern society.

33. Community relationships. — If you destroy the dam built by a colony of beavers, they set about its reconstruction, using the identical plan, method, and tools common to their species throughout all generations. Animal intelligence contains no quality that enables improvement beyond the inherited abilities or instincts of its kind. Herein lies the marked distinction between the highest type of animal and the lowest type of human intelligence.

Man possesses the ability to profit by the accomplishments of the past, to improve, and to develop. Upon this ability, the development of past civilizations has depended. Upon this same ability, the civilizations of the present and future are predicated. Out of this have grown community relationships established in ordered society upon the law of reason, supplanting the law of will, and ever-increasing in its benefits to all, with the growing understanding of the rights and worth of the individual member of society.

Coordinated action. — Coordinated group action has strength in so far as its members work together for the attainment of a common purpose — the subordination of self for the good of all. Only by helping others can we help ourselves.

"He profits most who serves best."

> In the development of her strength, wealth, and accomplishments, America is founded upon the establishment of successive communities bound together individually and collectively, by interdependent relationships. They are created and coordinated in home, school, church, and local self-government, as expressed in town meetings. Each individual member contributed his part to that greatest of all forces by which the character of the people of our Nation is sustained and developed — *public opinion.*

34. National relationships. — In the development of our colonies, the need of protection for person and property grew. We cooperated in the development of resources, and in exchanging products and labor in the creation of comforts and wealth. Our consolidated actions in resisting oppression and establishing rights created a national relationship. This binding of communities and States led to a federation designed for the welfare of all.

Articles of Confederation. — Under the Articles of Confederation, trade rivalries separated the new States from each other. There was an emphasis of State over National interests: One State lost its supply of cheap manufacturing material; industries suffered from want of coal, factories from lack of material, markets were limited;

economic barriers were set up, no cooperation existed, exclusiveness prevailed.

Constitution. — Grown now to a union of 48 States, working in a spirit of harmony and cooperation, restricted yet greatly benefited by our Constitution and statutes, we have come to be in terms of wealth, attainment, and influence, one of the outstanding nations of the world.

Under our Constitution, the departments of government are set up for the express purpose of coordination and cooperation for the general welfare of the Nation.

Interstate commerce. — Notwithstanding the sovereignty of each of the States composing our Union, great freedom is enjoyed as to residence, travel, trade, and property rights among their citizens which has developed an interstate commerce of tremendous volume and worth.

> Commerce among the States embraces navigation, communication, travel, the transit of persons, transmission of messages by telegraph. — *Justice Harlan.*

Railways, air transports, postal service, telephones, telegraph, radiograms help to unite the Nation by an exchange of goods or information, so that each citizen may know and profit by what the others are doing.

The Interstate Commerce Commission contributes to the development of "a more perfect union," which is an active association for cooperative effort. This commission touches the various interests of all of the people. Its benefits of regulations are in the interest of public necessities. It provides for a quick settlement of labor disputes affecting interstate trade and transportation, the control of which is lodged in the Federal Government.

35. International relationships. — In the development of those international relations which are in accord with the principles of interdependence, each nation must assume a larger responsibility and take a more active part in world affairs.

Due to the remarkable progress of civilization, isolation is no longer possible. International problems developing from everchanging economic and political conditions demand consideration and application of the principles of interdependent relationships as the means of securing the general welfare of mankind.

> I demand that the Nation do its duty and accept the responsibility that must go with greatness. — *Roosevelt.*

The State Department. — The State Department is the "friendly relations department" of our Government; by treaties and diplomatic negotiations, beneficent relationships with foreign counties are secured and insured, establishing a spirit of accord and amity without which it would not be possible to carry on our part in world affairs to the good of all concerned.

36. Beneficial to person and property. — The efficacy of our Constitution lies in the fact that it contains a statement of fundamental purposes relating to human associations and plan for their accomplishment, susceptible of such interpretation as to make them applicable to changing conditions.

Among the purposes set forth in the Preamble to the Constitution are "domestic tranquility" and "general welfare." The accomplishment of these purposes is based upon observance of the principles of interdependent relationships.

Law: Uniform acceptance and observance. — The security of persons and property is one of the inherent rights of mankind. It is guarded and guided by statutory laws, uniform in their restrictions and benefits, so that every citizen is fully protected in his rights.

Uniform laws are valuable in their benefits in proportion to uniform acceptance and observance. May a man have complete personal liberty? May a man do as he pleases? He may, provided he is not a member of organized society. To attempt such action as a citizen constitutes him an outlaw in such ratio as his independence interferes with the rights of others and breaks down the structure of government. All crime is, ignorantly or willfully, a violation of the principle of interdependent relationships.

Experience has revealed the necessity for united action to assure the greatest protection to the individual. Neither in person nor property will the individual find security without the assistance of his neighbor, community, State, and Nation. The higher the value we place upon human life and welfare, and the greater our accumulation of property, the more we must rely upon interdependent relationships based upon justice and inspired by mutual confidence and reciprocal endeavor.

37. Beneficial to production. — Industry is essentially the subjection of natural forces — the manipulation of natural material to the uses of mankind; it brings into action the worker, the engineer, the inventor, the organizer, the administrator, the combined energies of whom are liberated and set in motion by finance.

Accumulation of capital. — Thrift is the foundation stone of effective economic interdependence. The individual must practice frugality, engage in hard work, and acquire the habit of wise spending — so living within his means as to enable a saving of a portion of the product of his labor.

In industry, wealth is the product of saving; it is secured in part by the elimination of waste and the corresponding conservation of materials and labor practiced by both individuals and groups, and saving or the accumulation of capital is as much the duty of the employee as of the employer.

Relations between management and men. — To derive the greatest value from interdependent relationship between employer and employee, there must be created a spirit of good will and cooperation in which there is a recognition of mutual worth and mutual responsibility.

The atmosphere surrounding the relationship between management and men must eliminate fear, apprehension, and uncertainty. Only by the establishment of mutual understanding, confidence, and respect can effective cooperation and teamwork be secured. That employee renders best service who has an intelligent understanding of the relation of his part to the whole.

38. Results in progress. — Bound together by the ties of common interest and mutual benefits, society has advanced from:

> The crude hieroglyphic to the printed page.
>
> The smoke signal of the Indian to the radio.
>
> The tallow candle to the electric light.
>
> The hollowed log canoe to the *Leviathan*.
>
> The ox drawn prairie schooner to the airplane.

39. A Nation of specialists. — We are a Nation of specialists because experience has taught us that greater benefits will accrue to one and all through each individual learning to do one thing well.

> The physician looks after our health.
>
> The teacher gives instruction.
>
> The farmer grows the grain.
>
> The lawyer attends to legal matters.
>
> Others specialize in providing all the comforts and conveniences of home.

No one citizen builds his own house, manufactures the plumbing equipment, generates the electricity, constructs the heating plant, or provides the fuel for its operation. He does not pave the street, put in his own waterworks, provide police and fire protection, establish his own school, church, hospital, or theater.

40. Interdependence of capital, labor, and consumer. — Individual necessities, comforts, and conveniences as now enjoyed are the product of accumulated capital and labor, represented in modern organization, transportation, great factories, distant farms, tropical plantations, the trappers of the frozen northlands, the fishermen of the seas, and delivered daily to our homes by an army of tradesmen who administer to our wants and are in turn dependent upon us for their livelihood.

The telephone. — No better illustration of interdependence can be found than in the story of that all-necessary convenience, the telephone. It is difficult to imagine the diversified labor, the problems of transportation, the world-wide accumulation of materials, and the tremendous outlay of capital required in the manufacture of this marvelous instrument which receives and transmits the human voice regardless of distance.

Men toiling in the mica mines of India, in the platinum fields of the Ural Mountains, in the forests and jungles of far-off Asia, Africa, and South America, in the great forests of the Northwest, in the iron, copper, and lead mines, and the great steel works of the United States, produce the materials that go into the making of your telephone and the exchange controls.

The following raw materials, gathered literally from the four corners of the world, are used: Platinum, gold, silver, copper, zinc, iron, steel, tin, lead, aluminum, nickel, brass, rubber, mica, silk, cotton, asphalt, shellac, paper, carbon.

With the assembling of raw materials, and their fabrication in great factories into the completed instrument, there is added the work of organization and administration required in obtaining capital, franchises, building lines and conduits, installation of switchboards, and training personnel. Your telephone call to all points of the compass is made possible by these materials and the labor of nearly 400,000 employees in the United States alone.

41. Public utilities. — Public utilities corporations build great hydroelectric plants in one State for distribution of power to many. Coal, copper, iron ore are mined and transported to places of greatest advantage to industry. Railroad, telegraph, and telephone companies invest billions of dollars in properties and conduct their affairs to the benefit and profit of the Nation. Great dams are constructed and the desert lands of many States made fruitful by the vast irrigation systems created. Capital is consolidated and labor employed, farms enriched, cities built, and our citizens bound together in one cooperative, prosperous, happy union by the magic power of interdependent relationships.

Business. — Business, to ensure success, must keep in closest touch with the ever-changing affairs of social, economic, and political conditions. Vast sums of money are spent on new products, improved equipment, research laboratories, and inventions in creating new appetites and new markets.

42. Beneficial to peace. — In America, a degree of independence is developed out of which is born the idea in the minds of many that a citizen of the United States may be a law unto himself, retaining, however, the disposition to regulate the other fellow. If he does not like the law, he seeks a way to evade it, at the same time shouting vociferously over the increase of crime and the lessened influence of our courts. He demands the highest wages obtainable and complains at the prices he must pay for the product of his fellow laborer. He insists upon his right to independence and liberty, yet is ever ready to restrict such action on the part of others. That citizen who has not developed the spirit of cooperation, understanding, and tolerance is at war with his fellow man.

> The unity of good men is a basis on which the security of our internal peace and the establishment of our Government may safely rest. It will always prove an adequate rampart against the vicious and disorderly. — *Washington.*

Unselfishness. — Every American citizen must guard against the spirit of selfishness, the inordinate desire for material gain, the temptation to live beyond his means, and the tendency to find the easiest way to obtain the most in satisfying his constantly increasing wants.

Honesty — individual and collective, national and international — inspiring confidence wherein there is neither room for trickery nor unfair practices is the basis of the principle of interdependent relationships. Such honesty rests not so much upon legal rights as upon the Golden Rule.

43. Cosmopolitan character of population. — The United States, in her philosophy of self-determination, emphasizes the ideas and ideals of human rights and human associations. In the fulfillment of this policy, she opened wide her gates to the peoples of the earth, inviting them to share with her the blessings of liberty.

Somewhat less than half the racial stock of America's 108,000,000 white inhabitants is of British blood. Of the 95,000,000 whites, in 1920, 14,000,000 were born in foreign countries and 23,000,000 were of foreign or mixed parentage. There are 1,672,000 Germans, 1,600,000

Italians, 1,250,000 Russians, 500,000 Czechoslovakians, 465,000 Austrians, 370,000 Hungarians. There are 1,500,000 foreign born over 10 years of age unable to speak the English language. This foreign population supports over 1,000 newspapers published in 30 different languages.

There are no more untapped racial reserves.

Full privileges of citizenship. — The immigrant to America is particularly favored under the laws of the United States. Before the native-born youth can exercise the right of franchise, he must live under the influence of our system of Government, acquire his education, and enlarge it through associations and experience for a period of 21 years from his birth to his majority. It is possible for the immigrant (18 years or over), subject to certain restrictions to issuance of first papers, with little education, without that knowledge of our Government, association and experience, obtained only through years of residence, to have granted to him the full privileges of citizenship five years after his arrival.

Resultant duties. — In return for the opportunities and privileges established through her own sacrifices and paid for with the enormous exactions of treasure and human life, she expects — and has the right to demand that those who accept her hospitality shall respect her principles — that those who elect to live in the security and comfort of her homes and institutions shall give due honor and award full allegiance to her Constitution and shall in no instance, either by choice or through ignorant acquiescence, seek to despoil the land in which were bred freedom, equality, and opportunity.

The cosmopolitan character of the population of America emphasizes the burden which rests upon every citizen to become fully informed in the underlying

principles and ideals of our republican form of
Government.

Class consciousness. — Class consciousness and class
activity is the result largely of the intrusion of ideas of
government entirely outside of the fixed principles set
forth in our Constitution and should be no more tolerated
in our country than we would expect our principles, if
introduced by expatriated Americans, to be accepted by
another nation.

Immigrant not all problem. — The immigrant is not all
problem. He has been one of the outstanding assets in the
development of America. Slowly, but surely, there is being
assimilated and amalgamated in this country the bloods of
practically all nations, in the development of a racial stock
of exceptional worth in its vigor, ability, and character.

44. Our opportunity. — One of our greatest problems
is the education, assimilation, and amalgamation of these
various and numerous foreign groups into an
understanding, harmonious, loyal, and upstanding
American citizenship.

To this and succeeding generations is given the
opportunity to develop from our homogeneous character
an outstanding race expressive of the highest principles,
ideals, and traditions to which a God-loving,
humanity-loving, liberty-loving people can aspire. To
accomplish this great work, there must be a composition
of all differences which tend to create class consciousness
and class hatreds. Tolerance, born of knowledge,
understanding, respect, sympathy, and harmony,
engendered by the spirit of a common cause and purpose,
are essential in the interpretation of the principles of
interdependent relationships.

QUESTIONNAIRE
INTERDEPENDENT RELATIONSHIPS

Why did independence give way to interdependence? In what did it result?

Upon what has the development of civilization always depended?

What is the value of coordinate action?

State some of the principal causes that led to the creation of national relationships.

How did trade rivalry under the Articles of Confederation separate the new States from each other?

In what way was this situation changed by the Constitution?

How does interstate commerce assure a more perfect union?

How did railways, postal service, telephones, telegraph, and radio help to unite the Nation?

What is the attitude of the United States toward the problem of international relations?

What is the principal duty of the State Department?

In our complex civilization, may any individual live in complete independence?

Could any State maintain itself upon its own resources? Explain.

How are "domestic tranquility" and "general welfare" accomplished?

In what way does the individual find security in person and property?

What relations between management and men are essential to successful production?

What are some of the results in human progress that have been caused by the ties of common interest and mutual benefit?

What led the United States to become a nation of specialists?

Describe the interdependence of capital, labor, and consumer.

In what way does the telephone illustrate the principle of interdependence? Business? Public utilities?

What is the foundation of peace and prosperity?

What principal race stocks have contributed to American life? How?

What is the duty of America to our cosmopolitan population?

Is patriotism wholly selfish?

SECTION III

LESSON 3. — CHARACTER, THE GREATEST ASSET OF AMERICA

45. The greatest asset of America. — Diversity of opinion as to what is America's greatest asset creates a discussion which leads into every section and every activity of our country. Each individual is governed by the interest that lies closest to his heart.

The doctor declares: "The greatest asset of America is found in our medical schools, hospitals, and our great accomplishments in saving life and ensuring the health of our people, for without health there could be no other great achievement."

The teacher asserts: "Our common-school system, our colleges, universities, and our press constitute our greatest asset, for without education industry would stop and our Government disintegrate."

The captain of industry states: "Industry is our greatest asset. What would America be without New England, New York, Pittsburgh, Detroit, Chicago, and the thousands of other industrial centers giving employment to millions while they supply the needs of the world?"

Likewise, the inventor, the chemist, the scientist, each makes the claim that the fruit of his labor is the greatest asset of America, for what great things in America would have been possible without the creative genius?

The farmer insists that the doctor, the teacher, the industrialist, the scientist, and all the others would not get very far if he failed for a single season to provide the means for clothing and food — to him, the greatest asset of America.

They all are right; there are elements of greatness in all the varied endeavors of our country, the coordination of which has brought prosperity and wealth in such measure as to make us envied of all people.

46. Cooperation. — Forty-eight States, extended between the Atlantic and the Pacific, independent, self-governing Commonwealths, rich in resources, engaged in their own affairs; congested industrial areas of our great cities, pouring out their products to the enrichment of the Nation; millions of farmers providing food and clothing; teachers, preachers, merchants, laborers, lawyers, and clerks, contributing their part; all are working together in the spirit of cooperation.

47. Character. — What unites a people composed of all the racial stocks of the world? What composes our differences, harmonizes our relationships? What inspires confidence, ensures credit, and promotes organization? What, in the last analysis, guarantees protection of person and property, gives assurance of peace and prosperity, and inspires America to greater adventures and larger achievements?

The answer is not to be found in the sum of all her natural resources, factories, farms, homes, schools, hospitals, and churches. These are created by man and by man can be destroyed.

The security of our property, the continuation of our institutions, the increase of our possessions, and the perpetuity of the principles of individual rights, justice, and freedom, the observance of which has made America, lie in character — the greatest asset of America.

48. National character. — The ideals of the American colonists. — The national character of America is grounded in the Puritan stock of the early colonies. From these original settlers, numbering in 1640 a total of 26,000, there has descended to the present time nearly one fourth of our total population. Up to 50 years ago, their descendants and immigrants from the same racial stock composed over 80 per cent of our population.

The outstanding traits of their stalwart characters were defined in the commonplace affairs of their daily lives. They made no play for heroics, were not primarily seekers of fame nor fortune. Lovers of liberty, they boldly fought to maintain their rights. Their dominant trait was the worship of God, a God to be feared, yet a God of justice. A God who punished, yet a God who loved. Bigoted and narrow to the verge of superstition, intolerant of all faiths save their own, they built a character which to following generations will ever prove their richest heritage.

A stern will, born and bred of necessity, hard as the "stern and rock-bound coast" near which they lived, deep and cold as the seas that beat upon their rugged shores, they knew no compromise with duty — it must be done. No easy way was sought nor excuse accepted for duty unperformed.

Community life, church, and town meeting. — They established schools, churches, and town meetings, always dominated and often ruled with the iron rod of church authority. In time, bigoted religious intolerance gave way to religious liberty, yet not with the slightest change in the high standards of moral and spiritual rectitude required of every member of the community.

Possessing pride of race, proud of their ancestry, they inspired in the hearts of their children a reverence and respect for family and race, which left no room for lax conduct or easy habits. Severely disciplined within their homes, carefully supervised in their education, the children were taught the obligation of participation in community affairs and were obliged to submit to the severe restrictions imposed by their elders through the laws enacted by the local town meeting.

The restrictions of these laws and the severity of punishment imposed for the slightest infraction are cause for astonishment in these days of easy morals and lax law observance, yet their value as conducive to upright living, strict morals, and honest endeavor is strikingly evidenced by the pronounced influence of the New England community, the church and town meeting, in molding the national character of America.

> The New England town was founded for and grouped about the church, which was the clubhouse of the time. But the glory of the New England town was its town meeting, a combination of neighborhood, society, caucus, legislature, and council meeting. This was the most successful political institution of the time, served as a private school in debate, and a nursery of American statesmen. — *National Ideals Historically Traced* — *A. B. Hart.*

The pioneer spirit. — In defining the character of America, we find one trait so strong and pronounced as to manifest itself in every period and department of our national development — the "pioneer spirit."

Mixed motives inspired immigration to America. Regardless of why they came, the spirit of the pioneer seemed quickly to possess them with its urgent demand to go forth and conquer the wilderness.

In that spirit the New England pioneers, and those from the Middle Colonies and the South, peopled in succession the States beyond the Alleghenies and the Cumberland's, advancing by successive steps until they reached the boundaries of the continent.

The pioneer from New England and his cousins, the Scotch-Irish in Virginia and North Carolina, loved a struggle. To them, the wilderness held no terror too great nor hardship too severe to hold them back. Life was a joyful adventure and the dangers were enticing. Life held the stern duty of making provision for family and posterity. Life was work, and the great forests were there to be cleared. Life was full of promise; there were the vast free lands — theirs for the taking. Life was the gift of God and, never forgetting, they set the stamp of their God-fearing character upon each succeeding community, in school, church, and local government.

People from New England, the Middle Colonies, and the South, flowed together to form neighboring or joint communities, and thus varied the Colonial farmer type. This mixed population produced interesting combinations of local government; Michigan, settled largely by New England people, set up the town meeting; in Illinois, first reached by southerners, the county system was established in 1818, and later an option was allowed between town and county. — *National Ideals Historically Traced — A. B. Hart.*

Tenacity of purpose. — The very compelling forces of hardship, privation, danger, and isolation bred a spirit of unrestrained freedom, which has had a pronounced influence in forming our national character. Compelled to rely upon individual effort in providing and protecting his means of livelihood, the early American quickly acquired the knowledge of individual rights and the determination to maintain them. What was his, won by honest toil or by right of discovery, he was ever ready to defend against all odds.

Their tenacity in what they undertook has never been surpassed by any people, not even the Romans.

I remember that half the Plymouth colonists died the first winter, and that in the spring, when the long waiting *Mayflower* sailed again homeward, not one of the fainting survivors went with her — and I glory in that unflinching fortitude, * * * our stiffest muscle is limp and loose beside the unyielding grapple of their tough wills — *Doctor Storrs.*

This tenacity went far in possessing and saving to America the whole region west of the Mississippi River. The future welfare of the Nation, the preservation of representative government, and the principles for which it stands lie largely to-day within the hands of the citizens of the West, for into that section has traveled the center of our population, and there is to be found over one-half of the descendants of our Colonial forefathers.

Experimental self-government. — Our national character is emphasized in our ability to govern ourselves. Such ability did not develop over night; neither can it be acquired for the asking. No other nation has attained self-government in equal measure with the United States. The Colonies struggled 150 years before they had established a sufficient foundation to take the step that led to the "Great Experiment."

Our present form of government would never have been possible without this long period of preparation, involving study, experience, mistakes, and a growing measure of success, exemplified in the wise legislation inaugurated by several of the colonies, and in the increasing spirit of independence prior to the War of the Revolution. Success was made possible due to the collective fitness of the colonials for the task of self-government.

The colonial was "a good farmer, an excellent schoolmaster, a very respectable preacher, a capital lawyer, a sagacious physician, an able editor, a thriving merchant, a shrewd peddler, and a most industrious tradesman," able to comprehend the full measure of human associations. Hence, with these qualifications, when independence was won, a committee of chosen representatives called to the arduous task of revising the Articles of Confederation, found within themselves a collective knowledge which enabled them to produce that document, the Constitution of the United States, which, Mr. Gladstone said, "is the greatest piece of work ever struck off in a given time by the brain and purpose of man."

49. Individual character. — *Moral worth*. — In the discussion of moral worth, it is primarily true that we weigh and evaluate the actions of the individual. However, individual acts do not form a haphazard aggregate of unrelated deeds, for back of the act are dominant principles that assure a certain continuity in human action. With knowledge of the temperament and trend of mind of a given man, his action under given circumstances may be fairly predicted, due to the fact that behind the shifting play of emotions found in the mental life of everyone, there is a background of permanent emotional associations and processes which change slowly, if at all. This stable background of the moral life is character.

Deeds an expression of character. — Upon great impulse, one may commit an act foreign to his nature. However, in the long run of life, his deeds are an expression of his character. We base our estimate of character upon known performance; we catalogue the individual as good, bad, reliable, unstable, trustworthy, worthless. His worth to society is assessed. We judge what measure of reliance can be placed in him; how far he may be trusted; wherein lies his weakness, and wherein his greatest strength.

Public spirit. — The secret of the remarkable progress of America in the first 100 years of constitutional government lies in the fact that her public-spirited men were striving to put the best into government, not to take the most out of it.

No collective morality. — In the very nature of our Government, the responsibility for its social, economic, and political standards rests absolutely upon the character of its individual citizens. There can be no collective morality, integrity, honor, that is not the sum of the principles of the individuals of the community, State, or Nation. If the majority are mercenary, the character of the Nation will be ruthless. If the growing tendency to irreligious thought persists, the Nation will become irresponsible.

Desire for education and religion. — Desire is, perhaps, the greatest force in the determination of individual character. It overrules the handicaps of environment, poverty, and physical defects. It asks no favor of race, creed, or color. It has no determinate end. Its power is to ennoble or debase — "As a man thinketh in his heart, so is he."

The desire of our forefathers for education and religion, intensified with each succeeding generation by the ever-increasing facilities for intellectual development, has fixed the American character upon a high plane of moral worth and honorable attainment.

Knowledge is being extended with a rapidity and scope never before known in human history. By the magic of the facilities of modern communication, its voice is carried to the uttermost corners of the earth, challenging the present generation to newer and greater fields of adventure and achievement. The right to education is our heritage, established by our forefathers, guaranteed by the law of the land, enriched by our free institutions.

Notwithstanding this privilege, illiterates form a large proportion of our electorate. The National Education Association tells us that 4,300,000 illiterate citizens were qualified to vote in the last presidential election. Over 4,000,000 ignorant voters, unable to read any discussion of issues or candidates!

The last census disclosed that 1,400,000 children between the ages of 7 and 13 years were not in school during the period from September 1, 1919, to January 1, 1920.

Because of universal suffrage, the modern complexity of our national life, and the acknowledged principles of the right of private judgment — an open-mindedness receptive of the revelation of truth, a "thoughtful" citizenry is necessary.

> On the diffusion of education among the people rests the preservation and perpetuity of our free institutions. — *Daniel Webster.*

Foundation of character. — The character of the individual and the Nation is founded upon religion and education — which, united with that force we call "will," give to every individual the means for high attainment.

Submitting yourself to these impelling influences, resourcefulness and unconquerable energy take command. By their power, you win self-mastery. The joy of work becomes a reality. Labor is dignified by the pride of accomplishment. Obstacles and handicaps are but a challenge to greater effort. Discipline becomes self-imposed.

> Religion laid the foundations of our American Government. It neither seeks nor claims any justification for its existence save righteousness. It had its beginning, it found its inspiration, in the religious beliefs of the men who settled our country, made it an independent Nation, and maintained its institutions and laws. If it is to endure, it will be through the support of men of like mind and like character. — *President Coolidge.*

Daily performance necessary. — Expressed in terms of unselfish devotion to ideals, our attitude to others, our sense of responsibility, our willingness to give full service, loyal cooperation, our faithfulness to each other, and our reverence for religion, embodied in the daily performance of every task, "character" is the greatest asset of America.

50. Physical character. — *Great resources.* — Napoleon asked Talleyrand, "What is America?" To which reply was made, "It is a body without bones." An American adds: "The bones have been developed, and they are bones of steel."

Possessed of a raw continent, millions of square miles in area, composed of boundless prairies, vast forests, mighty rivers, great plains, and rugged mountains, containing fertile soil, rich natural resources in minerals, timber, and water power, the American, lacking tools, supplies, and capital, was forced by the very nature of his task and environment to a life of hard labor, long hours of toil, frugal living, and self-dependence with attending hardships and dangers. Out of these combined conditions was developed a type of hardy pioneer unequaled in the history of mankind.

Developed by pioneers. — In her commercial life, America has stepped boldly forth to the great task set before her. Slowly at first, groping her way along great rivers and through deep forests, she began the work of conquering the wilderness, which won as the fruit of her enterprise, first, the full possession of this great domain, and then, for her 300 years of toil, the greatest treasure house among the nations of the earth.

Visions become realities. — Forced to work by the very necessity of finding the means of existence, accepting danger and hardship, privation and suffering as a part of the task, America gave herself to creating material wealth.

Gaining strength and wisdom with succeeding years, America has built achievement upon achievement. No enterprise has been too great for her aggressive spirit. Her dreams and visions have become realities by the force of her will and the magic of her creative ability.

Spirit of improvement and invention. — Ever willing to adopt new ideas, to develop and improve, to tear down and rebuild, to scrap the machine of yesterday for the improved equipment of to-day, opportunity was never neglected to find a better way to do a larger business.

Success possible to every citizen. — Driven first by necessity, the joy of accomplishment became the spur to greater achievements. The way to comfort, to competency, to wealth was open and free to every citizen, limited alone by individual ability, courage, and determination. Out of great opportunity, with freedom to all, there has been bred a race of men and women of sterling character and outstanding independence.

51. Ethical character. — *Confidence.* — American business is based upon the character of its people. J. Pierpont Morgan used to say he banked more on a man's character than on his money. Character is the basis of confidence. Confidence is the basis of credit. Credit, above any other element, is the source of stability in commercial life. Our building industry, amounting to hundreds of millions annually, is dependent upon borrowed capital from the time of the first drafted plan to completion of each structure. The vast commercial enterprises of the United States are made possible by our system of credit based upon confidence in the integrity of the people.

The ethical character of our commercial relations is based upon respect for and confidence in the nobler things of life and the unfailing observance of business ethics.

High standards of commercial life. — America is a nation of corporations. Every enterprise of any consequence is incorporated. Founders or owners of a given business invite employee and public to share the risk and the profit. The workingman as a shareholder is rapidly becoming a capitalist; in number, they have increased to several million and their investments are assuming astonishing proportions. By this means, adjustments of differences between capital and labor are becoming easier as differences arise. The employer, in recognizing the employee as a fellow man and not as a commodity, opens the door to mutual understanding and square dealing.

As a stockholder, the employee feels the interest and responsibility of a partner. Greater attention is paid to the work, quality is improved, waste eliminated, and profits increased to the mutual advantage of all. The fact that labor is being less exploited and more fairly treated with each succeeding year is not only indicative of economic evolution but also a marked evidence of the high character common to our commercial life.

Spirit of cooperation and compromise. — One of the most encouraging signs of continued prosperity in America is the constantly growing tendency toward compromise and cooperation in the affairs of capital and labor, based upon mutual confidence. Such differences as are bound to arise are, as a rule, disposed of to the general good of all.

No class domination. — No class is permitted to dominate in America. Public opinion, which is always representative of public character, will not permit the assumption of power. Whether it be capital, labor, farmer, group, or section, public character in its dominant sense of fair dealing defeats the effort to acquire an unfair advantage.

Spirit of benevolence. — Nothing is more characteristic of modern American life than the pouring out of private wealth for public service. Nowhere are so many philanthropic agencies at work. *There is that in American democracy which creates the spirit of public service through gifts to the public.*

> **Copy Editor note:** This referral to "American democracy" has to be a typo. It was probably the word "decency" or something inserted by someone other than Harry Atwood. The term *"There is that"* only appears one time in the manual.

In respect to aid and contributions in world disasters, America is one of the first in the field of distress and one of the last to leave.

Not materialistic. — The Old World, looking upon the intense activity of this New World, seeing us ever engrossed in material affairs, with little time for leisure, even making hard work of our play in our overanxiety to win at any game, whether it be work or play, has scoffed at our lack of art, literature, and culture and called us a nation of dollar chasers.

Our justification for our so-called gross materiality lies in the fact that we were a new nation — new in a wilderness to be conquered; new in a land without homes, towns, or cities, without schools or churches, without transportation or communication. Under these circumstances, there was neither occasion nor opportunity to write music, paint pictures, or sculpture in marble. Our music was in the sweet, sonorous song of the mighty forests and the rushing streams; our pictures were painted daily in the mists of the morning and the waving fields of grain. Our monuments and memorials were carved from virgin forests, built in great cities, in rambling farmhouses set in emerald fields. We were kept too busy providing the necessities of life to find time for the finer accomplishments.

Now, lasting monuments depicting the strength, the majesty, and the beauty of our country are being erected; our large and well-kept parks are ornamented with beautiful sculptures; our colleges, universities, and institutions of music and art are comparable with those of any other part of the world; our public galleries and museums possess priceless works of art.

52. Political character. — *Diplomacy.* — America is slowly yet surely winning the confidence of the nations of the world. The desire to arbitrate rather than resort to armaments has distinguished America in her international policy, desiring to adjust all differences within the principles of justice and equity. Her commercial treaties are written in terms of square dealing. Backed by the guaranty of the American character, her obligations and her dollars are eagerly accepted wherever they may be offered.

QUESTIONNAIRE — CHARACTER,
THE GREATEST ASSET OF AMERICA

What is the result of the coordination of the "varied endeavors" of our country?

In what manner has the spirit of cooperation influenced the development of America?

Upon what does the perpetuity of our fundamental principles depend?

What are the main elements in the Puritan character?

What place in our early colonial life was occupied by the "town meeting"? What was its later influence?

State the chief characteristics of the pioneer.

Upon what is our estimate of character based?

What was the secret of our remarkable progress in the first 100 years of the constitutional Government?

What depends upon the character of our individual citizens?

Name several factors upon which our national character is based.

Why is religion an essential characteristic of the American people?

Are all American citizens educated? Explain.

How does public education affect American political institutions?

Why, under our form of government, is a "thoughtful" citizenry necessary?

In what manner is the gospel of hard work related to the American character?

Upon what is the ethical character of our commercial relations based?

To what is the success of our vast commercial enterprises due?

Why is the spirit of benevolence characteristic of America?

Is America materialistic? Explain.

In what is the political character of America expressed?

SECTION IV

LESSON 4. — GREAT AMERICANS AND THEIR ACHIEVEMENTS

53. The value of biography. |— The history of any nation, in its ideals and achievements, its motives and spirit, invariably reflects the character of its leaders. The stories of the lives and accomplishments of its great men are the windows through which is revealed the soul of the nation.

The biographies of the leaders of America should be carefully studied as the means of best understanding the controlling factors in the development of our country in any given period. In these stories are revealed the combat of minds, the clash of opinions, the cunning of politicians, the ruthlessness of self-seekers, and the saving forces of those dominant leaders who inspired the people to follow them in the establishment of the ideals out of which have been created the splendid achievements of our people.

In the lives of our great men are to be found the elements of righteousness, courage, justice, unselfish devotion to duty, self-reliance, initiative, and stubborn determination, the ingredients from which was brewed the virile, aggressive, and generous spirit of America.

With each succeeding period of our progress in government, industry, agriculture, education, medical science, we have had the leadership of men and women devoted to public service with little thought of personal gain.

In this spirit our Government was established. They who had power to assume control dared to commit that control to a free people, knowing that the ideals of liberty, justice, and individual right had been indelibly stamped upon the very souls of their countrymen.

In like spirit succeeding generations have responded to the call of their leaders for the preservation of our Nation. Creative and destructive forces are in eternal conflict. The experience of the past gives us wisdom to accomplish the tasks of the present.

Eternal vigilance is the price of liberty.

54. Fields of achievement. — The ideals and accomplishments of our great Americans were to establish a government that was fit to be entrusted with all the powers that a free people ought to delegate to any government as the safe and proper depository of national interests, controlled not by the passions but by the reason of the people, to develop the natural resources of the country, and to open up the way of opportunity to all.

However, great Americans have not confined their achievements to the field of government and protection of our institutions. Many of the great industries, much of medical science, communication, and transportation found first expression in the keen minds of our pioneers. In the biographies of these men are incidents as thrilling, full of daring, and productive of rich achievements as are revealed in the lives of the mighty army who conquered the wilderness and won for the United States in succession the Colonies, the Northwest Territory, Louisiana, the Oregon country, Texas, California, and the great Southwest.

55. George Washington. — This noble first citizen of America is an outstanding character in the history of our country. From his early youth, he demonstrated those qualities of leadership which, with the experience gained in his great achievements, made him the dominant personality of his time.

Military leadership. — At the early age of 23 years, he was placed in command of the Virginia Rangers. He became the hero of General Braddock's ill-fated campaign against the French and Indians. After General Braddock's failure to accept his advice, which caused his (General Braddock's) death and the defeat of his troops, it was the superior ability of Washington which saved the British from rout and possible annihilation. As commander in chief of the Continental Army, he took command of a disorganized, undisciplined yet loyal body of raw provincials. Ragged and starved, half frozen, and poorly equipped, by the force of his character, he brought them to a condition of training and discipline that gave final success to the Colonial cause.

By the charm and strength of his personality, he won the admiration and enthusiastic support of the great German general, Von Steuben; the brilliant Frenchman, Count de La Fayette; and the gallant Pole, Kosciusko.

Political leadership. — The conclusion of the war found General Washington so exalted in the hearts of his countrymen as to make him the virtual ruler of the new nation, created largely through his military genius and indomitable will. Foregoing all personal ambitions other than that of molding a free people into an enduring nation, he gave himself with equal faithfulness to the work of peace and orderly government.

Serving without pay in all his public career, his life of unselfish devotion rightfully won for him the title of "Father of His Country." When charged by an unfriendly Congress with usurpation of power, he replied: "A character to lose, an estate to forfeit, the inestimable blessings of liberty at stake, and a life devoted, must be my excuse."

Inspired by the influence of his character and his qualities of statesmanship, such men as Hamilton, Jefferson, Madison, Randolph, King, Marshall, Monroe, and the venerable Franklin addressed themselves with him to the task of constructing a new government, which in the following generations was destined to become an ever-growing memorial to their wisdom and patriotic devotion to the ideals and rights of humanity.

Farewell address. — The wisdom, sagacity, and vision of Washington gave the United Colonies a republican rather than a democratic form of government. In the almost inspired words of his "Farewell Address" — in the framing of which he undoubtedly had the assistance of Alexander Hamilton and James Madison, two of the finest minds of that period — he gave counsels concerning the pitfalls which have destroyed other popular governments of history. As far as these counsels have been observed, the Nation has enjoyed peace, prosperity, and happiness.

The Nation's gratitude. — George Washington, born February 22, 1732, died September 14, 1799. Within the scope of his 67 years, he was surveyor, farmer, soldier, statesman, commander in chief of the Continental Army, and twice President of the United States of America.

More than to any other, we owe our everlasting debt of gratitude to George Washington for American independence and the Constitution of the United States.

56. Benjamin Franklin. — Benjamin Franklin manifested the qualities characteristic of the American. Genius he possessed, but it was the genius of hard work. He was a self-made man. At the age of 17 years, he came from Massachusetts to Philadelphia, which became his lifelong residence.

Printer, publisher, philanthropist. — A journeyman printer by trade, he ultimately became the author and printer of Poor Richard's Almanac, a publication of homely philosophy which contains many gems of wisdom and good advice as applicable to-day as in his time.

Franklin was identified with the Pennsylvania Gazette. He founded the Saturday Evening Post, the University of

Pennsylvania, and the Philadelphia Public Library. He was the first postmaster general of the Colonies.

Scientist. — With all these activities, he still found time to devote to science. The flash of lightning in a thunderstorm caused him to wonder rather than to fear. In it, he recognized a mightily force; his philosophic mind concluded that in some way, the flash of lightning (electricity) could be controlled and brought into the service of mankind. It pays to think. Creative minds, as exemplified in Franklin, rather than manual labor, have produced the great achievements of all time.

Political philosopher. — Benjamin Franklin was too busy to be idle. Absorbed with the affairs and welfare of the Colonies, he proposed in 1754 that the Colonies be formed into a Union. Franklin believed that had this proposition been accepted, a separation from the British Empire would never have taken place. Twenty years later, a call for a general congress of the Colonies was issued by Virginia, at the instigation of Franklin, and held in Philadelphia in May, 1774.

Benjamin Franklin took an active part in framing the Declaration of Independence, of which he was one of the signers.

Diplomat. — Two years later, he went to France, where, in fur cap and homespun clothing, he, the typical American commoner, created a wave of enthusiasm which won the French to the cause of the Colonies.

Member of Constitutional Convention. — At the age of 81 years, this old young hearted philosopher took a most prominent part in the deliberations of the constitutional convention held in Philadelphia from May to September, 1787. His wisdom and counsel often prevailed in those long and stormy sessions. *His love of country and faith in democracy gave him a vision of the future greatness of America that few in his time possessed.*

"...to determine this Question I appealed to Our own experience, and to the experience of mankind from the earliest Ages to the present, & have ventured to say that in the long review there cannot be found a single instance of any Nation's enjoying Peace, Liberty, and safety, under a Democracy;—That whilst a Democracy had existed in any Nation, it had ever been attended with violent feuds, parties, & civil disorders or wars which (a democracy not having sufficient energy in its Nature to suppress) soon produced general Anarchy the constant prelude to Tyranny and despotism;"

57. John Marshall. — The life work of John Marshall is intimately blended with the Constitution. He ranked high as a soldier, legislator, diplomat, historian, and statesman. As a jurist and magistrate, he ranks first. For 34 years, he served as Chief Justice of the Supreme Court of the United States, leaving a record for constructive results in the Government of the United States second only to that of Washington.

Soldier. — He began the study of law at the age of 18 years, but soon left his studies to enter the Revolutionary Army. His experiences, with their heroisms and hardships, "broadened his views and quickened his insight in governmental questions." He said, "I entered the Army a Virginian and left it an American."

Member of assembly. — After the war, he was elected a member of the Virginia Assembly. During his remarkable career, he served in the legislature for eight sessions. He continually emphasized his conviction that for efficiency, a strong central government was necessary.

Ratification of the Constitution. — As a member of the State convention, in 1788, which met to discuss the ratification of the Constitution of the United States, through the power of his convincing logic, the ratification of the Constitution was accomplished over the determined opposition of its enemies.

Member of Congress. — At the urgent request of Washington, he became a candidate and was elected to Congress, where he became the greatest debater on constitutional questions.

Interpretation of the Constitution. — In 1829, through his wisdom and moderation, he did much to prevent radical changes in the State constitution of Virginia, thwarting the attempts of politicians against the independence of the judiciary. Because of his exceptional understanding of the philosophy of the Constitution of the United States, his counsel was of prime importance.

His deep convictions and illuminating arguments contained in his decisions concerning constitutional questions, at a period when the powers of the Constitution were ill defined, were of inestimable value in the formation of a well-organized Federal Government. "He made the Constitution live. He imparted to it the breath of immortality. Its vigorous life at the present time is due mainly to the wise interpretation he gave to its provisions during his term of office."

> The most notable products of Marshall's unprecedented judicial career may be summed up under two heads. In the first place, he established the supremacy of Federal law within the entire circle of its jurisdiction, no matter whether it was opposed by the Congress or by a State legislature in the form of unconstitutional enactments, or by the President giving "instructions not warranted by law"; or by State supreme courts attempting to resist the mandates of the Supreme Court; or by the governors of States attempting to resist such mandates; in the second place, in defining the character of "the American Constitution." — *Origin and Growth of the American Constitution — Hannis Taylor.*

58. Thomas Jefferson. — By reason of his ability as a thinker and speaker, Thomas Jefferson quickly gained a place of leadership, first in Virginia, then in the Colonies, where he was constantly employed in fighting oppressive British regulations and interference in the affairs of his country. Staunch in his defense of the rights of the people, he caused Virginia to pass many laws of a revolutionary character, among which was the abrogation of the rights of nobility, entailed estates, and the absolute right of religious liberty.

Declaration of Independence. — He was a member of that famous group which, upon call of the resolution proposed by Richard Harry Lee, wrote the Declaration of Independence. Although the youngest, his dominant personality and exceptional ability caused him to be chosen chairman of that committee, which included such stalwarts as John Adams, Benjamin Franklin, Roger Sherman, and Robert R. Livingston. The instrument practically as written by Jefferson was unanimously adopted to become for all time one of the immortal documents relating to human rights and self-government.

President of the United States. — In the trying days during and following the Revolutionary War, Thomas Jefferson was a member of the Continental Congress, Governor of Virginia, ambassador to France, succeeding Franklin, and recalled to become Secretary of State in President Washington's Cabinet, where he bitterly opposed the policy of Alexander Hamilton in his endeavor to extend the powers of government over the people.

On a platform based upon his ideas and policies, he was elected the third President of the United States as a Democratic-Republican over his opponent, who as a Federalist, espoused the principles of Hamilton.

Louisiana Purchase. — During the first years of his two terms as President, he completed the negotiations with France for the purchase of the vast domain, over 900,000 square miles, lying west of the Mississippi River and east of the Rocky Mountains, known as "the Louisiana

Territory." The purchase price of $15,000,000 was, at that time, considered exorbitant and created much adverse criticism in which Jefferson was denounced as an "imperialist," and as having forsaken his democratic principles. The reasons for this action on his part were that he saw the advantage of gaining control of the Mississippi River and the port of New Orleans, and that by this purchase the United States would be left unhampered by foreign countries in developing her republican form of government.

Achievements. — The outstanding events of his public life are to be found in (1) the writing of the Declaration of Independence; (2) enactment of the statute for religious liberty; (3) founding the University of Virginia; and (4) the purchase of the Louisiana Territory.

59. Daniel Webster. — Daniel Webster belongs to the first generation of Americans who knew no other form of government than that established by the Federal Constitution. So intimately is his name associated with that great document that he has become known to history as the greatest expounder of the Constitution.

Tampering with the Constitution. — When but 20 years old, he delivered an address which contained the following:

> The experience of all ages will bear us out in saying that alterations of political systems are always attended with a greater or less degree of danger. The politician that undertakes to improve a constitution with as little thought as a farmer sets about mending his plow is no master of his trade. If the Constitution be a systematic one * * * its parts are so necessarily connected that an alteration in one will work an alteration in all, and the cobbler, however pure and honest his intentions, will in the end find that what came to his hands a fair, lovely fabric goes from them a miserable piece of patchwork * * *.

Representative government. — As a further caution against a pronounced tendency, he declared:

The true definition of despotism is government without law. It may exist in the hands of many as well as one. Rebellions are despotisms, factions are despotisms, loose democracies are despotisms. These are a thousand times more dreadful than the concentration of all power in the hands of a single tyrant. The despotism of one man is like the thunderbolt which falls here and there, scorching and consuming the individual on whom it lights; but popular commotion, the despotism of the mob, is like an earthquake, which in one moment swallows up everything. It is the excellence of our Government that it is placed in a proper medium between these two extremes that it is equally distant from mobs and from thrones.

Webster clearly understood our representative form of government and the importance of avoiding the dangerous extremes of either hereditary (autocratic) government or direct (democratic) government.

Reply to Hayne. — Webster's replies to Hayne in the United States Senate are considered as the greatest debate that has ever occurred in any legislative body in the history of the world. His second reply began with the following words:

Mr. President, when the mariner has been tossed for many days in thick weather and on an unknown sea, he naturally avails himself of the first pause in the storm, the earliest glance of the sun, to take his latitude and ascertain how far the elements have driven him from his true course. Let us imitate this prudence, and before we float farther on the waves of this debate, refer to the point from which we departed that we may at least be able to conjecture where we now are.

This indicates a wholesome state of mind with which to approach important discussions concerning the philosophy of our Government, as expressed in the Constitution. Before we drift farther toward direct action and socialistic tendencies, we should return in study and thought to the work of the men who wrote the

Constitution and ascertain how far we are departing from the course therein laid down.

60. Abraham. Lincoln. — George Washington gave us the Union. Abraham Lincoln saved the Union.

Log cabins were common in this country 100 years ago. It was not a log cabin that gave distinction to Abraham Lincoln, although he was born in the poorest of such cabins on February 12, 1809.

Limited education. — His honors were not conferred upon him because of a university education. Two short terms in a Kentucky school, followed by three in Indiana, less than a year in all, does not give much foundation for scholastic attainments.

Handicaps. — To study the life of Lincoln makes one almost believe God purposely placed every conceivable handicap upon him just to prove his staying qualities, and to set an example in purpose, principle, and perseverance. This would act as an inspiration for young and old possessed of the same ambition and endurance, the vision and character necessary to achieve success.

Abraham Lincoln was homely, yet he possessed the beauty of soul dedicated to relieving the burdens and sorrows of humanity.

He was a rail splitter. In his rugged physical strength, he was as gentle as a woman.

His was a lowly birth, yet "his spirit is the richest legacy of the United States."

Lawyer. — He was a "saddlebag" lawyer, yet, with a copy of Blackstone, a Webster's Dictionary, and the fundamental law of God and human rights in his heart and head, he won his way to the respectful consideration of all opponents.

With his sense of humor and ability as a story teller, there was in him a super sense of justice, and he often fitted a story to emphasize a truth that otherwise might have been forgotten.

Preservation of the Union. — "A house divided against itself cannot stand." Upon that issue — the preservation of the Union — Abraham Lincoln was elected President of the United States. Tolerant with all who opposed, kind to all who hated, charitable to those who denounced, he held firmly to the single purpose of saving the Union, in the belief that in union, only could our Nation endure.

The beauty of diction, the reverence, sympathy and love, the magnanimity and charity, and the vision of the worth of the price paid for the preservation of our Union, as set forth in his Gettysburg speech, will make him acclaimed after all other orators are forgotten.

The nation incarnate. — He was the nation incarnate. In all its struggles, its doubts, its agony, and in the solemn days of victory Abraham Lincoln lived alone for his country.

No one man has ever rendered greater service nor paid a greater price for faithful performance. As he has given us a rich legacy in his spirit and example, so he has left us a great responsibility —

> That we highly resolve that these dead shall not have died in vain; that this Nation, under God, shall have a new birth of freedom; and that government of the people, by the people, and for the people shall not perish from the earth.

61. The winning of the West. — In a brief space of time, 50 years, was accomplished the stupendous task, entitled by President Roosevelt "the winning of the West," an accomplishment made possible by the sturdy character of the men and women who so fearlessly and laboriously carried on once they set their faces toward the golden West.

Accustomed to frugality and hard labor, inured to hardships and privation, stern in self-discipline and faith, mighty in determination and self-reliance, they not only left to posterity an inheritance of fertile land, virgin forests, great water resources, and untold mineral wealth, but, greater than the sum of all material gain, they passed on to this and succeeding generations the principles and traditions of independence, liberty, and justice, an example of the worth of clean living, high purpose, and great faith that should be an inspiration to every loyal American.

In the original grant of charter to the several Colonies by Great Britain, the western limits were practically undefined. Several of the Colonies claimed territory extending westward as far as the Mississippi River and north of the Ohio to the Great Lakes.

Northwest Territory. — In the compromises made, composing the differences between the Colonies, it was agreed to define the western boundaries of such Colonies to more restricted areas, dedicating the disputed territory to the United States, to be known as the "Northwest Territory," which at the time was occupied by French and British trading posts.

This area included what are now the States of Illinois, Indiana, Ohio, Michigan, and Wisconsin. All territory lying west of the Mississippi River and east of the Rocky Mountains, from the Gulf of Mexico to an undetermined northern limit, was then a possession of Spain known as the "Louisiana Territory," transferred by Spain to France and then sold in 1803 to the United States.

With the exception of a few venturesome spirits who found their way across the mountains south of the Ohio River and as far west as the Mississippi, this land of ours was an unknown wilderness to the settlers of the Colonies. Alive with deer, buffalo, and small game, rich in timber, fertile of soil, watered by numberless rivers and lakes, America at the close of the War of the Revolution still awaited discovery.

Slow development. — The thrilling story of the winning of the West is a series of events accomplished not by military force but rather by the efforts of a host of hardy pioneers who, with indomitable fortitude and incredible labor, won in succession the swamps, rolling prairies, forests, plains, rugged mountains, and the fruitful Pacific slope.

> Copy Editor note: He just cannot bring himself to say, "Defeat the Indians." In 1928, they were still considered savages.

No single individual dominated this vast domain. It was the rank and file who conquered in this battle of the wilderness. Its conquest was not quickly accomplished. As in all great movements, leadership was developed, with here and there a man who became identified with some particular period or section.

Daniel Boone. — A native of North Carolina, born and developed under conditions that gave him physical strength and endurance beyond the average, courage, daring, and self-reliance, he was peculiarly fitted for what he declared to be the mission of his life — "ordained of God to settle the wilderness." He was the highest type of wilderness explorer. Living to the age of 86, he will continue to live throughout the annals of our history as an outstanding type of the earliest American. He exemplified in his life the value of clean living, high principles, and hard labor.

Settlement of Kentucky. — Undaunted by the unknown dangers of great swamps and forests, matching wits and woodcraft with the roving bands of hostile Indians, he led the first group of settlers across the Blue Ridge Mountains into the rich country of Kentucky.

Here, amidst untold hardships, privations, and danger, there was set up the beginning of what has grown to be a mighty State, rich in natural resources and richer still in the treasure of its manhood and womanhood, descendants of the sturdy stock of Daniel Boone and those who followed him. These hardy pioneers bred into the succeeding generations that strength of purpose, endurance, initiative, and determination which has contributed so much to the richness and virility of American character.

62. George Rogers Clark. — Capt. George Rogers Clark saved the settlers in Kentucky from massacre by the Indians and was the hero of the conquest of the Northwest Territory, now represented by Illinois, Indiana, Ohio, Michigan, and Wisconsin.

Military expeditions. — He led his small force of less than 200 men against the French outposts of southern Illinois. With their capture, he turned his attention to the British garrison at Fort Sackville on the Wabash River at Vincennes, Ind.

In the capture of this fort Captain Clark and his sturdy band accomplished one of the most difficult marches in military history. Crossing the "drowned lands" of southern Illinois in the month of February, 1779, they carried on through water oftentimes above their waists, without provisions or supplies other than that carried upon their backs. Through a wilderness untraveled and unknown by white men, this small band of backwoodsmen took the British by surprise, demanded and received the unconditional surrender of the garrison. By this remarkable exploit, America was forever rid of foreign domination, and title to this region was given to the United States.

His monument. — Capt. George Rogers Clark was among the greatest of the forefathers of the mid-West. By the inspiration of his spirit, fortitude, and courage, this handful of men acquired possession of this inland empire of America. By acts of heroism, serving without pay, and assuming the debts contracted in this campaign, Captain Clark magnified his devotion to his country. The memorial to his self-sacrificing service is not to be found in tablets or statues of bronze, but rather in the great Commonwealths that now comprise this territory — the heart of America.

63. Lewis and Clark. — In May of 1804, Captains Meriwether Lewis and William Clark proceeded to St. Louis, Mo., in obedience to the following order issued by President Jefferson by authority of Congress:

> Go up the Missouri to its sources; find out, if possible, the fountains of the Mississippi and the true position of the Lake of the Woods: cross the stony mountains, and having found the nearest river flowing into the Pacific, go down it to the sea.

The expedition. — Outfitting in St. Louis, Captain Lewis and Captain Clark, with four sergeants and twenty-three privates of the Regular Army, and an Indian interpreter, began the long, tedious journey up the swift current of the Missouri, reaching its headwaters approximately one year later. Crossing the Rocky Mountains, through the Bitter Root Range, they found the Clearwater River. Proceeding down its course through exceedingly rough country to the Snake River, in what is now Idaho, they continued on to the Northwest to the junction of the Snake with the lordly Columbia.

Launching their canoes upon the broad reaches of this most beautiful stream in October, 1805, they drifted down to the Pacific Ocean, reaching their destination on November 7, one month later. Returning from there to St. Louis, with their surveys and maps of the regions explored, they completed the required journey in a little over two years' time.

Claim of United States to territory established. — How little was known of the great domain secured to the United States in the purchase of the Louisiana Territory is revealed in part by the wording of the President's order. How much was learned and its importance to the Nation was contained in part in the report those two intrepid Army officers gave upon their return. The most important result obtained was the firm establishment of the claim of the United States by overland exploration, its first claim being made through the earlier discovery of this north Pacific country by Captain Robert Gray, of Boston, who sailed his ship from the Pacific Ocean up a great river in 1792, naming it the Columbia, in honor of the three hundredth anniversary of the discovery of America by Columbus.

The new country. — The Lewis and Clark expedition gave the people their first idea of the vast area, enormous natural resources, and grandeur of the Pacific Northwest. They were the forerunners of what soon became a mighty host of emigrants into the land of the setting sun

64. Rev. Marcus Whitman. — Thirty years after the Lewis and Clark expedition, Rev. Marcus Whitman packed all his earthly possessions in a wagon and, with his bride, trekked across the plains and mountains, over what became known as the Oregon Trail, to the Walla Walla country as a missionary to the Indians.

Impressed with the beauty and richness of the country, he seemed to have lost sight of his special mission, as seven years later, he took the trail back to civilization, there to urge his countrymen to follow him in the possession of this new land.

Western emigration. — Acting as guide for this band of emigrants, recruited largely in New England, he led them ever westward in the all but impossible journey of nearly 4,000 miles. The story of the hardships and perils, the labor, sickness, and starvation, the fight with Indians and nature, serves again to prove the sturdiness, self-reliance, and courage of the pioneers of America.

Sterling qualities of racial stock. — Every advancing step in the progress of our Nation emphasizes the sterling qualities of the racial stock that, handed down to succeeding generations, has given the urge and the will to do, the fruits of which are to-day enjoyed by a prosperous and happy posterity.

Boundary adjustment. — These men and women, who so bravely followed Whitman over the Oregon Trail, saved that great country to the United States. The cry in 1846 was "The British must go — The whole of Oregon or none — 54 40 or fight." In the spirit of fair play and justice, the differences with Great Britain were adjusted, the boundaries were fixed, and another great step in the expansion and settlement of our Nation was accomplished.

65. Gen. John C. Frémont. — As a junior officer of the United States Army, at the age of 29 years, Frémont was designated by the Secretary of War to explore a route from western Missouri to the "South Pass."

Exploration of the Southwest. — In accomplishing his mission, he followed the Arkansas River to its source in the Rocky Mountains. On a later expedition, he made his way through Utah to the Great Salt Lake and then through the deserts of Nevada and across the Sierra Nevada, where he found his journey leading through the mammoth trees and along the roaring torrents of the California country, reaching the Mexican city of Monterey, some 130 miles south of San Francisco on the Pacific Ocean.

Mexican War. — Through exercise of diplomacy, he was able to remain in this vicinity until after the outbreak of the Mexican War, when he headed a revolt against that Government and freed the territory of California from Mexican authority, becoming the governor of the territory which was ceded to the United States by treaty following the conclusion of the war with Mexico.

A contemporary. — Contemporary with Frémont, another brilliant young Army officer, Colonel Kearney (afterwards brigadier general), fought his way across the plains of Texas to Santa Fe, N. Mex., and after its capture, continued across the deserts of New Mexico, Arizona, and southern California to a union of his small army with Frémont in California.

Territorial acquisition. — As a result of the splendid work of these men coupled with the success of Generals Scott and Taylor in old Mexico, there was added to the domain of the United States the last of the great southwestern area, a territory of nearly 1,000,000 square miles, a section of our country which within one year thereafter became the goal of the adventurous spirits of the world due to the discovery of fabulous gold deposits along many of the water courses flowing to the Pacific Ocean from the western slopes of the mountains bordering eastern California.

66. Eli Whitney, a pioneer of modern industry. — *Invention of cotton gin.* — A school-teacher from Massachusetts living in Georgia in 1793 invented a machine called the cotton gin, by use of which a negro could easily clean 300 pounds of cotton a day, demonstrating thereby, as no previous invention had done, the value of machinery in replacing or augmenting manual labor. The whole question of cotton production and cotton manufacture was changed through the use of this invention.

Previous to the invention of the cotton gin, cotton yarns were spun and woven into cloth by hand in private homes. Necessarily, by this slow method of manufacture, but small quantities of cotton were used.

Development of cotton industry. — So rapid was the development of the industry, stimulated by this new "gin," that within the next 20 years, exports of cotton to Liverpool increased tenfold.

As a result of this invention, a cotton factory was erected in Massachusetts to produce cloth like that made in England. Here was constructed the first loom operated by water power in America. In 1814 there was built at Waltham, Mass., the first cotton mill in the world, in which the raw material direct from a Whitney cotton gin was spun into thread, woven into cloth, and printed with colors, all under one roof.

Influence on country. — The production of cotton was stimulated and made one of the leading industries of the country. Cotton exports enormously increased; allied industries developed; communities grew rapidly into cities.

The invention of the cotton gin created unforeseen social, economic, arid political conditions; it largely put a stop to the discussion of slavery; the southern planters and northern manufacturers of cotton found it to their mutual interest to keep the negro in bondage, since by his labor they were rapidly growing rich.

Due to climatic conditions, the manufacture of cotton goods was carried to New England, thus opening a new channel of employment, causing in following years a radical change in the nationality of the citizens of these Northern States.

Interchangeability of mechanical parts. — While Whitney was the inventor of the cotton gin, because of the theft of his model and tools from the shed in which he conducted his experiments, he was not enabled to perfect his invention.

He instituted the interchangeability of parts which has greatly influenced modern industry. In 1798 he secured a contract from the Government for the manufacture of firearms, being "the first to effect the division of labor by which each part was made separately." It was from this invention that he made his fortune.

67. Robert Fulton, a pioneer of steam navigation. — It is proper and fitting to designate Robert Fulton as the pioneer of modern transportation by reason of his success in driving the Clermont, in the year 1807, against the current of the Hudson River from New York City to Albany.

Other inventors. — It is true that no less than eight men had at various times and places propelled boats by steam power prior to this accomplishment by Robert Fulton, yet none of them carried out their experiments to a successful issue.

Fulton's success was largely due to his cleverness and ingenuity coupled with the fortunate circumstance of a partnership formed with Robert Livingston, a man of wealth, also interested in solving the problem of steam navigation.

Legislative grant. — Livingston was so sure of final success through his own various experiments as to induce the Legislature of the State of New York to pass a bill granting exclusive right to navigate the waters of that State by steam power upon condition that a boat of 20 tons be driven by steam at a minimum speed of 4 miles an hour against the current of the Hudson, this feat to be accomplished within one year from the date of grant. He failed in his effort. Later he was appointed minister from the United States to France.

The "submarine." — In 1803, while in Paris, Fulton demonstrated his "submarine" in the River Seine. Encouraged by the success of this experiment, Fulton and

Livingston ordered a steam engine from Watt & Boulton in England, to be shipped to America, where Fulton found it on his return in 1806.

The "Clermont." — In the following year, the *Clermont* was built and launched in East River. Its successful trip opened the way to a complete revolution of water transportation. Within the next few years, so rapid was the adoption of this new method of travel, steamboats came into use upon the principal rivers and the Great Lakes, rendering splendid assistance in establishing easy communication between distant sections of our country traversed by the great waterways.

Progress in water transportation. — To fully appreciate the value of the contribution made by Fulton and Livingston to the economic development and enrichment of America, one has only to review the remarkable progress made in water transportation, contrasting the present accomplishments with those 100 years ago.

Through his vision, patience, and persistence, he found success where others had failed, and in so doing, opened the way to the rapid development of this mighty agency in the advance of civilization.

68. Samuel F. B. Morse, a pioneer of modern communication. — Without our present facilities of communication, modern civilization could not continue. Deprived of telegraph, telephone, and radio, the wheels of industry would be stopped, and the economic welfare of nations destroyed. We cannot too greatly emphasize this benefaction conferred upon all people through the accomplishment of Samuel Morse and the brilliant men who followed him with improvements upon his basic invention, the telegraph.

Opening of the Erie Canal. — Morse trained himself to think. Of all the thousands whose attention was engaged by the opening of the Erie Canal in 1825, he alone caught the significance of the passage of time in relaying the message heralding that event. The signal was delivered by cannon placed at intervals between Buffalo and New York City, the successive reports of which, conveyed from one emplacement to the next, consumed one and a half hours of time in delivering the message a distance of 500 miles.

Invention of the telegraph. — Reason and logic compelled him to believe that electricity made to travel many miles over a copper wire in an instant of time could, by some method, be interrupted in its passage so as to produce certain signals susceptible of interpretation.

Busy in his profession as an artist in London, Italy, France, and at home, the idea of the control of electricity ever persisted in his mind. With the passage of years, his patience was rewarded with the invention of a crude telegraphic instrument and a system of dot and dash signals to be used therewith. Forming a partnership with Alfred Vail, they labored together in the perfection of the device until their funds were exhausted.

Appropriation from Congress. — Undismayed, their persistent appeal to Congress for assistance was finally rewarded with an appropriation of $30,000 for the erection of a telegraph line a distance of 40 miles between the cities of Baltimore and Washington. With the completion of its construction, on the morning of May 24, 1844, in the presence of the chief officers of the Government, in the Supreme Court room of the Capitol, Professor Morse, operating the key of his instrument, successfully transmitted to the wonder of all present that first and memorable message, "What hath God wrought?"

Improvement and amplification. — Morse was a man of vision. He predicted the day when telegraph lines would span the earth and bridge the seas, yet even his far-seeing mind could never have encompassed the stupendous results which have come from his creation as a rich boon to all mankind.

Men great in scientific accomplishments have followed with improvements and amplifications upon his invention. Alexander Bell and associates applied his principle in perfecting the telephone; Thomas Edison improved the technique as telegraph operator and inventor, following his own powers of deduction into still broader fields. Marconi and others enriched his creative efforts in the field of wireless communication. Each passing year witnesses' other improvements and accomplishment, all a living testimonial to Samuel Morse, the man of vision, who, standing apart from the crowd, sold himself to a great idea, and persisted against all odds until his efforts were crowned with success.

69. Capt. John Ericsson, pioneer of the modern battleship. — John Ericsson, a native of Sweden, directed his inventive genius to improvements in steam navigation. He claimed the invention of the screw propeller but was unable to prove priority.

Coming to the United States in 1839, he built the first screw propeller warship, the *Princeton*. This was the first steamship ever constructed with her boilers and engines below the water line, and was the beginning of the steam marine of the world.

The "Monitor." — Ericsson would probably have remained unknown to the nation at large had it not been for his achievement during the Civil War. Using the revolving turret patents of Theodore Ruggles Timby, he combined a structure with all machinery below the water line, leaving the turrets alone exposed to attack. This small vessel, known as the *Monitor,* called in derision "The Yankee Cheese Box," in its victory over the *Merrimac,* made Ericsson famous in a day.

The navy and merchant marine. — This caused a revolution in naval development among the world powers, increasing the effectiveness of fighting ships, thereby greatly strengthening the offensive and defensive forces of nations in proportion to their naval tonnage.

Through the genius of John Ericsson, the modern navy and merchant marine has become one of the greatest factors in the development and security of nations.

70. Major Walter Reed, conqueror of yellow fever. — Major Walter Reed, a surgeon in the United States Army, conducted a long series of experiments in Cuba and discovered the source of yellow fever to be in the Stegomyia mosquito. The dream of his youth had been to be permitted to alleviate, in some degree, the sufferings of humanity, and all his efforts, without a thought of self, were spent in striving toward this goal. Within a few months after this discovery, Habana, which had been ravaged by this disease for more than 150 years, was cleared of the disease.

71. Major General William C. Gorgas, conqueror of malaria. — Through the efforts of Major General William C. Gorgas, who was in command of the medical and sanitary organizations of the United States Army in Panama, this pestiferous district was converted into a healthy region. The French enterprise on the Isthmus of Panama was completely wrecked by the fevers common to that region; 75 per cent of the employees from France died from the disease within a few months after they had landed on the Isthmus. As a result of the intensive efforts of Doctor Gorgas, the situation was conquered, and Panama has become one of the healthiest spots on the continent.

James L. Tippins

QUESTIONNAIRE — GREAT AMERICANS
AND THEIR ACHIEVEMENTS

What is the value of biography?

What, in general, were the ideals and accomplishments of the great Americans?

Describe briefly the influence of George Washington on the Nation.

Who was Benjamin Franklin and in what way did he influence the development of the country?

In what way did John Marshall contribute to national welfare?

What advantages did Thomas Jefferson secure for the United States by making the Louisiana Purchase?

Against what modern movements did Daniel Webster counsel?

To what principal task did Abraham Lincoln set himself?

Who was Daniel Boone, and what did he accomplish?

As the result of the expedition of Captain George Rogers Clark in the Northwest Territory, what States were added to the Union?

What was achieved by the expedition of Lewis and Clark?

How did the efforts of the Rev. Marcus Whitman terminate in reference to the Oregon country?

Who was General John C. Frémont, and of what value were his services?

Who were the real conquerors of the West?

What were the main steps in our national development accomplished by far seeing American statesmen?

What principal changes were brought about by Whitney's invention of the cotton gin?

What was Whitney's greatest invention? Why?

What were the principal contributions of Robert Fulton in modern development?

Who invented the telegraph?

Who improved and amplified this invention?

For what are we indebted to Captain John Ericsson?

Who made the discovery that stopped the ravages of yellow fever the world over?

Who eradicated tropical anemia and malignant malaria from Panama?

SECTION V

LESSON 5. — ECONOMIC DEVELOPMENT OF AMERICA

72. **The colonial spirit.** — Three hundred years ago, America was a wilderness. Her total white population consisted of a few hundred men, women, and children, established in several small communities along the Atlantic seaboard. For the most part, they were a God-fearing people, led to America by the vision of a new land in which they could work out ideals and visions inspired by their deep religious convictions. Along with these groups were others of more worldly persuasion, who came in the spirit of adventure or to escape political conditions, which, in the changing reign of the rulers of England, made their move advisable.

73. Colonists largely representative. — As a whole, the colonists were largely representative of the life, thought, and aspirations of that period; they were not supermen and women any more than they were of the vicious type. They were moved by the impulses common to humanity, chief of which is always that of self-preservation.

74. A continent to conquer. — Here, they found a vast and unknown continent in the possession of roving tribes of Indians; a wilderness of great forests, mighty rivers, and boundless prairies. Theirs for the taking, if they possessed the ability and courage to conquer the all but insurmountable obstacles and dangers.

Limited facilities. — Forced by lack of any other means than those contained in hand and brain; lacking all facilities of communication, transportation, or manufacture, other than such contrivances as the sailing vessel, the ax, spinning wheel, wooden plow, and flint-lock rifle, their progress in the first 150 years was necessarily slow and restricted.

Chief pursuits, agriculture and seafaring. — The colonists labored under the burden of heavy restrictions imposed by the mother country, which prevented the establishment of home industries. As their first occupation, they engaged in tilling the soil that they might have food and clothing.

During her first 150 years of existence, America grew to be a people of some 3,000,000 souls and was forced to confine her development to agriculture and seafaring pursuits. Building up a seafaring trade, she transported the raw material of the new land to England, France, Holland, and Spain, there to be exchanged for the necessities of life not produced by their own handicraft.

75. The federation of the colonies. — Industrial progress came with the establishment of the new Nation, "The United States of America." Lacking capital, other than that of character, courage, and concentrated labor, the bankrupt colonies were welded into a union of action, which has led our Nation by successive stages to its present attainments, the marvel and wonder of modern time.

76. Encouraged by constitutional provisions. — In the Government set up under the Constitution, provision was made for a freedom of action which gives full play to every citizen in the exercise of his rights and powers. The wisdom of the law of our land is emphasized with each passing year. The remarkable economic development of America is based upon the liberties and restrictions granted as the equal right of all her citizens. Outstanding among these provisions are —

The money clause. — The money clause establishes credit through the sole power vested in the Federal Government to coin money, incur national obligations through issue of bonds or notes of indebtedness, establishment of our national bank, and later our Federal reserve bank system, forbidding any State from incurring financial obligations with foreign powers or other States.

The post-office clause. — The post office clause, through which communication is regulated between the States and with the world at large, is a duty alone of the National Government. In this clause are found the rules and regulations governing mail, telegraph and telephone lines, and the radio. Strict regulations hold all accountable for matter transmitted by mail, as to its truthful or fraudulent character; rates are fixed by the Government with equal application to all.

The commerce clause. — The commerce clause set up an agency of exceptional worth by reason of the freedom granted in interstate traffic, the elimination of barriers, duties, or restrictions which might otherwise be created in exchange, sale, and shipment from State to State. Citizens of any State have equal rights as citizens of the United States, subject only to such local laws as apply to all citizens of the State within which business is transacted.

The taxing clause. — The taxing clause permits taxes to be levied for the requirements of government only; such taxes to be uniform in application and subject to revision as necessity governs.

The naturalization clause. — The naturalization clause establishes one class of citizens only; with equality to all and privilege to none. Under this and the immigration acts, our Nation is assured a strength and unity of purpose and action, and an equality of citizenship that could not otherwise be attained.

Fixed terms of office. — Fixed terms of office: Our system of government by which definite terms of office are assured gives stability to business in the fact that in no crisis can an administration be overthrown in a day, through dissolution of Congress or the resignation of the Cabinet. Parties may rise and fall without serious effect upon our economic life.

77. Free land and opportunity. — Other important factors in our economic development were free land and diversified natural resources. In these, America has been particularly blessed. Lack of capital prevented none from making progress in America. For the first 250 years, the immigrant to our shores knew that the door of opportunity was wide open. Landing with barely enough money to pay transportation to this chosen destination, and with no hindrance other than that of being a stranger in a new country, both land and employment were to be had for the asking.

Westward Ho! — Through the liberality of our Government and the vast and areas open for settlement, there was established and developed the largest and richest agricultural territory now under cultivation in the world. For nearly 100 years following the War of Independence, the cry was "Westward Ho!" By families and by groups, the creaking ox-drawn schooners wended their way slowly toward the setting sun. Driving the Indians and wild game before them, they cut the forest, broke the sod, planted, harvested, built home, school, church, and town, preparing the way for the next step in our progress — the railroad.

78. Influence of the Civil War. — Before any great railroad development had taken place, the peaceful life of our country was interrupted by the Civil War. It is questionable if that struggle, with its frightful loss of life and treasure, would ever have taken place had railroads been constructed linking the North and the South. In 1860 there were only some 30,000 miles of railroad in America, nearly all of which ran east and west, by reason of the fact that our great rivers flow from the north to the south, and our railroads could not then compete with river transportation. In 1860 no railroad was built farther west than the Mississippi River. West of that stream, the country was almost entirely given over to the great herds of buffalo and roving Indians.

With the close of the Civil War, the impetus given industry by the necessity of making war materials, the development of steel, and a growing appreciation of the value of rail transportation caused a marked advance in our economic life. The acquaintance of masses of men from every section of the country and the close ties formed by their association through the war added its force to the awakening of a new era.

79. Capital control. — Capital saw great opportunity for profit through development of our vast natural resources. Foreign capital was attracted. Combinations were formed. These groups were able to obtain concessions and rights, quickly developing a power of control over industry which placed in the hands of a comparative few the economic life of America.

Need for cheap labor. — With capital consolidated, only labor was required for this exploitation of our natural resources. America was too vast in area and too small in population to furnish the labor. By then-existing immigration laws, the doors were open — the world might enter. Capital needed labor, and it must be cheap labor.

The new immigration. — "The man with the hoe" was invited and urged to find in free America his great opportunity. He came by thousands, then tens and hundreds of thousands.

The former class of immigrant had come to America to take up land and become farmers and builders of homes and communities. They were followed by the thousands who worked in the noise and sweat of our great steel mills, in our coal mines, and in the factories, which quickly built up within our cities large congested areas, with great sections almost entirely composed of single nationalities. Labor was exploited, voted, worked, or left unemployed.

80. Citizen control. — Following the war with Spain in 1898, a change was inaugurated. Led by far seeing men who recognized the danger to our free Government in the increasing power of capital, the people developed a system of control through Congress which broke or checked its combinations. Industry had greatly developed during this period. Wealth had been amassed as never before. Yet the economic life of America had suffered — equality of opportunity was largely restricted and classes with intense class hatred were created.

81. Adaptation to abnormal conditions. — In 1917, there came a national emergency. One class alone — the "American citizen" — took precedence. America astounded the world with her ability to adapt herself to abnormal conditions, converting her peacetime factories and equipment to war time requirements.

82. Labor advancement. — During the World War, the wage earner learned to put his excess money into Liberty bonds. He caught the idea of investment, acquired the habit of systematic saving, discovered the strength that lies in consolidating the small savings of the many. He began to understand the meaning of capital, lost his fear of it, and found a way to have a part in its benefits.

83. Mass production and high wages. — The conclusion of the war found America committed to mass production, mass cooperation, and mass saving. These were some of the blessings that accrued out of the hell of war. Industry awakened to the astonishing fact that high wages to labor increased rather than diminished profits by the simple process of increasing the buying power of millions of employed.

84. Steady employment. — Industry learned the value of steady employment. It sought ways of regulating production to give work the year around. Seasonal employment ate up savings, weakened buying power, destroyed credit, and increased cost of production caused by idle equipment and accumulated stocks.

85. Intensive efforts of industry. — Industry set up research bureaus, stimulated inventors, chemists, and scientists to greater efforts in a search for better machines and methods, the elimination of waste in materials, and in developing by-products therefrom. Through these intensive efforts, production per manpower has been largely increased, new products created, markets enlarged, and industry stimulated.

86. The creditor Nation. — In the earlier history of American industry, foreign capital was invested by millions of dollars in our great enterprises. We were a debtor Nation. To-day we are the creditor Nation.

87. Production the basis of wealth and wages. — There is no actual wealth in materials, metal, or money until they are adjusted to the use, needs, or wants of mankind. Production is the basis of wealth.

In no other country do wages approach the sum paid the individual workman of America. The contributing factors to this highly satisfactory situation are summed up in the word "production." American production per man power ranges from two and one-half to thirty times that of other nations.

88. Mechanized industry. — Industry in America is mechanized and specialized to a degree not approached by any other country. Our automatic labor-saving and power-driven machinery is the wonder of Europe. Our mass production, made possible by special machinery and highly trained operators, astounds the world with its magnitude, quality, and low cost.

89. Higher self-appreciation. — Modern methods of industry discipline the lazy, wasteful, and disloyal workmen; speed up production; work out short cuts; improve quality; and eliminate waste; thereby contributing largely to lower costs through greater efficiency. At the same time, there is engendered a higher appreciation in the employee of his worth to himself, his employer, and his country.

90. Employee becomes employer. — A keener sense of pride awakens ambition, a quickened intellect inspires study, a broader view of life reveals opportunity, creates hew desires expressed in higher living standards, and a rapidly growing participation in industry as a partner through purchase of stock in different enterprises. Through quickened intelligence and systematic saving, the employee of to-day becomes the employer of to-morrow. At a dinner in New York given in 1927 to a group of British workers investigating American industry, every American captain of industry present save one came up from the overall stage.

91. High standards of living. — Human needs are few by comparison with human wants. Were it not for ever increasing desires for the comforts, conveniences, and luxuries of life, modern industry would be unable to sustain itself. Civilization is the result of human demands, the combination of spiritual and material aspirations. In no other nation have these aspirations been so fully satisfied.

The standard of living established by any group or nation is based upon the distribution of wealth. The closer together we bring the extremes of wealth and poverty, the higher the attainments and general welfare of the people.

Ability to purchase. — Power of consumption is based upon the ability to purchase and pay for the desired commodities. In America, the employee receives 72 per cent and the employer 28 per cent of the income of industry, constituting a range of wealth distribution which fixes our living standards at the highest point known in the world.

92. Is America worth saving? — The remarkable development of American industry has proven beneficial to all — not only to employer and employee, but also to the world.

America has amassed unbelievable wealth which is being spent for the good of mankind. In its large range of distribution, it has fixed our standards of living at the highest point known to civilization.

We may therefore answer — Yes! America is well worth saving!

QUESTIONNAIRE
ECONOMIC DEVELOPMENT OF AMERICA

What facilities for economic development were available to the early colonists?

What were their chief pursuits? Explain.

In what manner did the Government, set up under the Constitution, encourage economic development?

Name other important factors in our economic development.

Describe the impetus given to our industries by the Civil War.

What led to the demand for "cheap labor"?

In what way did "the new immigration" compare with the colonists?

How did the people control industry?

Explain America's adaptation to abnormal conditions during the World War.

Describe the benefits to labor through high wages and steady employment.

How do research bureau's aid industry?

What is meant by the "creditor nation"?

Upon what does prosperity depend? Explain.

Is mechanized industry beneficial to the people?

How do modern methods of industry affect labor?

What are some of the benefits of mechanized industry?

How do America's standards of living compare with those of other nations?

What is the range of wealth distribution in America?

What conditions and qualities have made possible the creation of the great wealth of America?

What can you do to assist in the further economic development of America?

Is America worth saving? Why?

SECTION VI

LESSON 6. — INDIVIDUAL INITIATIVE

93. **Mankind a mass of individual ego.** — Psychology and social science have discovered that mankind is made up of a mass of individual ego, each revealing similar characteristics of instincts, idiosyncrasies, and

manifestations of selfish interests — in the control of which his intelligence has developed forms of government.

From earliest childhood, self-assertion, self-determination, and self-preservation manifest themselves.

It is human nature for the strong to take advantage of the weak, whether it be strength of body, strength of mind, or strength of a group; that group may be a minority in numbers, yet all-powerful by reason of the forces under its control.

The chief purpose of government is that of controlling this instinct and directing it into channels through which society will gain the greatest benefit.

94. Two forms of government. — One form of government gives the State the supreme control and places all its citizens upon a common level of "equal condition"; the other recognizes the rights of the individual as greater than the government, and emphasizes the superiority of "equality of opportunity" in contrast with "equality of condition."

95. Collectivistic government. — "Equality of condition." — In this system of government, stress is laid upon the proposition that "all men are created equal," meaning that no man has a right to that which is denied to another; that any system of government failing to recognize and conform to this "ideal" is wrong, and therefore an enemy of society and a foe of mankind.

The ignorant, illiterate, physically and mentally deficient, the lazy, improvident, and reckless have equal right with the alert, aggressive, busy, educated, high-minded, orderly citizen who aspires to the best and is willing to pay the price of attainment through self-discipline, hard work, and careful management.

It is not in human nature to recognize "equality of condition" except to acquire a personal advantage. One may be willing to divide another's property with the third and fourth individual providing the share remaining to him is something more than he formerly possessed.

Denial of personal rights. — "Collectivism" is the denial of personal rights. The State (community) becomes the chief concern of all. It claims that the "law of equality," once applied, would destroy every human desire for individual dominance, making society safe, content, comfortable, and happy.

This "ideal" is to be accomplished by the application of force under the direction of leaders, in the selection of whom the people will have little or no choice. It is necessary, at first, to enforce the will of community interests until the people become educated and submissive to the new order.

Denied all personal rights, "collectivism" gives its "instructions" where to live, where to work, what to do, what to think, and what to say, for the State is the law.

Confiscation of private property. — "Collectivism" declares that the possession of property has developed protection of property through governments, courts, police power, and public opinion, making it difficult for one to acquire private property except by work. Private property must be abolished so that all will live on a plane of "equal condition." As a matter of fact, however, "human nature" will see to it that the "equal condition" will very quickly become an equal condition of misery, want, and discontent.

Religion outlawed. — The collectivistic government proceeds against "imperialism" by outlawing the church. The church at the behest of capital "fed the people the opium of religion," making them willing slaves to do the will of their capitalistic masters. In the interest of the new order, there must be left no place for religion, lest the people gain courage to throw off the yoke of their new found freedom.

Abolition of the family. — With personal rights, private property, and the church abolished, to make subjection complete, "the state" declares that in pure "collectivism" there can be no family ties, for children, like all other property, are an asset of the community and must be robbed of family love and obligation as a necessary step to loyalty to the state. Marriage may be practiced if conscience insists, but is not demanded in the interest of the new society, for with the abolishment of personal rights, private property, church, and home, society no longer possesses a moral, ethical, or spiritual code.

"Socialism" kills. — The doctrine of "socialism" is "collectivism." It tears down the social structure, weakens individual responsibility by subjection to or reliance upon the state in all material, social, and political matters. It compels the thought that at his best, man is no better than the worst; he loses his self-respect and his keener sense of moral and ethical values. Ambition is nullified by restriction of choice in occupation and reward of attainment. Initiative, the very backbone of all progress, is smothered in the morass of impersonal service, mass servility, and mob inertia.

"Socialism" aims to save individuals from the difficulties or hardships of the struggle for existence and the competition of life through calling upon the state to carry the burden for them.

"Equality of condition," the ruling law of "collectivism," is the death knell alike to individual liberty, justice, and progress through the destruction of individual and national character.

When the citizens of a nation, seeking comforts and pleasures, find no joy or satisfaction in hard work, the years of that nation are numbered. Free bread and the circus marked the declining days of Rome. A surfeit of food, clothes, comfortable homes, and much time for idleness can easily become the first step to the overthrow of civilization.

96.   Individualistic government. — *"Equality of opportunity"* — "Equality of opportunity" carries with it the absolute right of every man to keep what is his own. There can be no confiscation of property without due process of law and just recompense to the rightful owner. Upon this foundation have been based most of the great accomplishments of the past as well as assurance for still greater achievements.

Right to private property. — Each citizen enjoys the right to private property. Granted the privilege of working for one's self ambition is fired, initiative is encouraged, labor is not restricted, and the hard thinker and hard worker gets the reward denied the lazy and indifferent, creating thereby classes, caste, poverty, and wealth.

Economic freedom. — The individualistic form of government, promotes and guards the individual amid the difficulties and hardships of his struggle for existence and in the competitions of life.

The workman is protected because the nation needs his labor and the employer is protected because the nation needs his industry.

The productive power of free initiative has full play and a sure reward. Under its protection, he finds joy and satisfaction in the fruits of his labor. There is incentive to invention, improvement, and the establishment of families and homes.

Political rights. — It protects the citizen in his personal freedom. Equal political rights are assured. He has a voice in the Government which is "of the people, for the people, and by the people."

When a people are free to undertake things and take advantage of the opportunities open to them, wealth, character, and national strength are developed.

Protection to home and family. — The social unit of civilization is the family. Under this form of government, the institution of marriage and the rights of childhood are respected, the home and the family are protected, and womanhood is inviolable.

Respect for religion. — The "individualistic" form of government believes in the exercise of religious freedom and shows tolerance toward and respect for all religious beliefs.

The American Government rests upon the deep religious convictions of her people. If it is to continue, it will be through unceasing respect for and confidence in the nobler things of life.

97. An American institution. — In the governments of the Old World, conditions which built up a fixed caste system and created an impassable barrier between certain groups of society gave exceptional advantages to the favored and denied to the masses all but a bare existence.

The early settlers of America, who came to escape the oppression of this order of society, at first incorporated into the local governments of the Colonies the policy of religious intolerance and class rule. It required 150 years of local experiment in colonial government before the inalienable rights of mankind were sufficiently understood and evaluated to develop the necessary public opinion and power to change the prevailing form of "State" government to that of a "Republican" form, under which "equality of opportunity" became an American institution.

"Individualism," an experiment in government, was unknown prior to the independence of America, and has proved its worth by its marked achievements.

It tolerated no restriction, recognized no exceptions, and demanded that the son of the farmer or frontiersman have the same opportunity as the son of the merchant prince or land-owning aristocrat.

98. Constitutional guaranties. — The American citizen knows that he and his children may attain any goal to which intelligence, courage, and ability may lead. No overlord will ever bother or hinder their advancement. No succession to power or property is vested in titles of nobility to be transmitted through succeeding generations to favored families. The rich of to-day may be the wage earners of to-morrow, while the story of the rise of the exceedingly poor to affluence and power is as common as it is true.

The young American's future depends upon himself. He may inherit a fortune; his sense and ability alone will enable him to keep it. He may be born in the cabin of the miner or the shack of the mountaineer, yet if within him there burns the unquenchable fires of ambition, courage, and indomitable will, there are none who may stop him on the road to success.

> No person shall * * * be deprived of life, liberty, or property without due process of law; nor shall private property be taken for public use without just compensation.
> — *Constitution of the United States.*

99. Aristocracy of brains. — The only aristocracy that America will ever recognize is that of "brains" — "the tools to him who can handle them." The tribute in honor and the reward in wealth accorded to brains in this land of opportunity are not equaled in any other country. Brains ask for no "equality of condition" and want only "equality of opportunity."

100. The four "I's." — Socially, economically, politically, the world is rapidly changing, and, in its evolution, it requires for its leadership men and women of individuality, independence, initiative, and intelligence.

Individuality. — Under the guaranties given by the Constitution, there has been developed in the American character a striking individuality, which stamps him an American wherever he may be found. It is that quality which inspired him to the conquest of the great American wilderness and the development of her resources. The urge of individuality has driven him in every undertaking not only for pecuniary reward but for the equal reward of stamping his achievement with his own personality. This distinctive bearing of the American commands attention and wins the confidence of all.

Conscious of his own strength, he asks no other favor than equal opportunity. When he marries, he seeks no dower with his bride. He accepts his place in life with dignity born in the consciousness of his own power to better it. Be it ever so humble, his home is marked with his personality. His children bear the impress of his character, giving assurance that life can contain no difficulties too great for them to master. His is the consciousness of the free born, whether born in the crowded tenement of a congested city, the lonely prairie home of a western farmer, or within the sumptuous palace of a millionaire. Imbued with the spirit of the Nation, he stands upon his own feet and gladly enlists as a soldier in the battle of life.

Independence. — The American is the personification of independence. He asks no favors of government or men. He demands his rights and is always ready to uphold them. He has cultivated the habit of self-reliance and is ready to undertake any legitimate enterprise, which, in his judgment, has a reasonable chance of success. Resourceful and unafraid, he has ventured into every field of endeavor, cheerfully paying the cost of his failure and as cheerfully sharing with others the rewards of his success. In the spirit of independence America has won her way to leadership in times of peace, and in times of war to a place of honor and respect among the nations.

Initiative. — Out of independence has grown a force of individual initiative which has made our great achievements possible. Initiative might well be termed the generator from which has come the power for all our accomplishments. Tradition looks always to that which is old in habits, customs, culture, government, institutions, families, and structures. Initiative is forever putting off the old and putting on the new. It is the mother of creative genius, expressed in science and invention.

Without initiative, civilization would first stagnate, then fall rapidly into dissolution.

In no community in the world is freedom of initiative enjoyed as fully as in America. Government, laws, customs, and traditions operate to enhance that freedom.

Intelligence. — So far, our minds have grasped each successive problem and found so far, the means of meeting each added complexity of modern civilization. With multiplied wants and ever-expanding fields of endeavor, the demand for intelligence increases. Machines are taking the place of hands, increasing production, shortening hours of labor, eliminating the exhaustion of toil, giving more time to self-betterment, recuperation, and recreation.

Markets become world-wide, competition grows keener, international affairs demand care and diplomacy; nations are awakening; the magic of science in transportation and communication has made us largely a family of nations with divergent aspirations, varied needs, and growing demands for self-expression.

101. The price of success. |— The price of success, whether of individual or nation, is found in work, education, and ideals.

Work. — The world grows busier with each passing year. Its machinery is never idle. Its burdens are too great to be encumbered with dead weight. Backward individuals and backward nations will surely be crushed beneath the Juggernaut we call civilization, unless they take a more active and intelligent part in its affairs.

There is more and greater work to be done with each succeeding generation. The achievements of individuals in the past are a challenge to the youth of to-day. There are still further fields of exploration, adventure, and accomplishment, and a multitude of past achievements to be perfected. Every man possessed of the will to work finds his opportunity awaiting him.

Education. — Education, he must have. The time is past when hope of success can be offered to the ignorant. With each succeeding year, the necessity for special accomplishments and particular fitness is more pronounced. Science has so far advanced as to become broken into many divisions, each requiring special training. Applied to every branch of government, industry, and even society, the demand is for education, that intelligence may be developed and applied to its full capacity; for in no other way may progress be assured, and progress is the purpose of life.

Ideals. — Work and education are not sufficient to equip either the individual or nation for the accomplishment of the purposes of life. There must also be the inspiration and governing force of ideals. Without ideals, there can be no lasting achievements. Without ideals, there can be neither understanding, tolerance, justice, nor brotherhood between individuals or nations. Without high ideals, there can be no worth-while aspirations, no true nobility of character, no spirit of unselfish service, all of which are essential to real progress.

102. The citizen's privilege. — Emerson said, "Hitch your wagon to a star." The citizen should demand of himself and for himself the best that life affords, and devote his energies in an ever-growing measure to public service, for the real joy of life is service to our fellow men.

This is the land of "equality of opportunity." The citizen alone can determine the measure of his participation in freedom's field. What he does and how he does it will be dependent upon his will to work, the thoroughness of his education, and the quality of his ideals.

We are a country of 118,000,000 people, speaking one language, having an enormous consuming power and an adequate transportation system for prompt distribution. We are not restricted within our wide limits by artificial barriers. We produce where it is most advantageous and distribute to the consumer where he may live. Here in the East, we may eat the apples and use the timber from the Northwest, and the Pacific slope may buy cotton cloth from the Carolinas and motors from Detroit. Nowhere in the world does there exist so large, so varied, and so unrestricted a market as the United States.

> There is a force underlying these factors and one which to me is all important. I mean the initiative and energy of the American people. We are willing to work. We have that divine restlessness which will not permit us to accept things as they are but drives us to find something better. We are constantly improving our machinery, our methods, ourselves. Here no man accepts the level into which he has been born as fixing his status for life. Ability is quickly recognized; to rise is easy. * * * There is movement, not fixation, in our life in America. — *Andrew W. Mellon, Secretary of the Treasury.*

QUESTIONNAIRE
INDIVIDUAL INITIATIVE

What is the chief purpose of government?

What is the fundamental principle of "collectivists" government? Explain.

Describe four of the principal elements of "collectivism."

What is the general effect of "socialism"? Describe.

Should the government provide the means of livelihood? State reasons.

Has any government the right to restrict the exercise of the power of individual initiative?

What is the fundamental principle of "individualistic" government?

Name and describe five of the principal elements of "individualistic" government.

Explain the origin of "individualism" as an experiment in government.

In what manner does the Constitution guarantee political, economic, and social freedom for the American citizen?

Name four characteristics of the American character.

What determines the success either of an individual or a nation?

In what way are high ideals essential to real progress?

State the synonym for "America."

What responsibility does freedom of initiative place upon the American citizen?

SECTION VII

LESSON 7. — LIBERTY AND INDEPENDENCE

103. Historical background. — The historical background of liberty and independence is the story of the human race in every stage of its development and in every corner of the earth. It is told in the ages-old pyramids of Egypt, built upon the backs of human slaves; in the philosophies of Plato and Socrates; and uncovered in the catacombs of Rome. In the German forests, it was planted deep in the hearts of Saxon and Norman, and there given its first real semblance of form.

England, in the days of the Saxon and Norman conquest, in the time of Cromwell and Elizabeth, laid a still broader foundation upon which to build the structure of self-government.

Slowly there was evolved an appreciation of government incorporation of the rights of individuals into fixed laws or practices. Yet there remained the iron heel of government to crush those whose demand for independence and liberty exceeded that granted by the will of the ruling King or Parliament.

104. Slow development of necessary knowledge. — It remained, however, a work still to be accomplished at the time of the first settlements in America, where in the next 150 years, slow progress was to be made in developing the necessary knowledge upon which liberty and independence could safely rest.

105. The Declaration of Independence, a protest. — The Declaration of Independence was a protest against the abridgment of such rights as the colonists claimed as subjects of the British Crown. Their anger was directed against Parliament rather than the King because restrictions were placed by law upon the colonists which were not imposed upon citizens of Great Britain residing in the mother country. These operated solely for the benefit of the long-established home government and institutions. Spurred by the spirit of independence engendered through the bitter experiences and necessary self-reliance required in their century-and-a-half battle to conquer the American wilderness, and fired by the indignities and injustice to which they had long been compelled to submit, they threw off the yoke of oppression and set up a government that would forever guard them against tyranny, however it might seek to impose its will.

> When in the Course of human events, it becomes necessary for one people to dissolve the political bands which have connected them with another, and to assume among the powers of the earth, the separate and equal station to which the Laws of Nature and of Nature's God entitled them, a decent respect to the opinions of mankind requires that they should declare the causes which impel them to the separation. | — We hold these truths to be self-evident, that all men are created equal, that they are endowed by their Creator with certain unalienable Rights, that among these are Life, Liberty and the pursuit of Happiness.

> And for the support of this Declaration, with a firm reliance on the protection of divine Providence, we mutually pledge to each other our Lives, our Fortunes and our sacred Honor — *Declaration of Independence.*

> No man sought or wished for more than to defend his own. None hoped to plunder or spoil * * * and we all know that it could not have lived a single day under any well-founded imputation of passion. — *Webster.*

Independence of the Colonies. — The American Colonies did not become free and independent until they were strong enough to throw off the yoke of the oppressor; strong enough to set up and control their own Government through the voice of the people; strong enough to protect and defend their country from aggression whether from within or without.

Its enemies. — The "enemies within" who would make the Declaration of Independence a mockery play one group of Americans against another. They fan the flames of prejudice. They magnify fancied evils of injustices to the ignorant. They distort its language to suit their own ends so cleverly that many of the less informed follow them in the name of Americanism.

Its survival. — Every American citizen must be constantly on guard if the principles set forth in the Declaration of Independence are to survive.

106. Liberty defined. — There are two kinds of liberty — absolute liberty: That of the savage, in which any individual may act as he pleases; and civil liberty: That of a civilized community in which human actions are regulated by law for the good of all — subject only to such restraints as a solemn and tolerant judgment determines to be essential.

> Political liberty is no other than natural liberty so far restrained by human laws and no further, as is necessary and expedient for the general advantage of the public. — *Blackstone.*

Liberty does not free the people from the necessity for control, but it places a heavy burden of responsibility upon the individual for self-control. It is not license to do as one pleases. Through developed "intelligence," man has power to control his baser and more selfish instincts, compelling their exercise and restriction in the interest of society.

Minority control exercises its will until such time as general intelligence becomes sufficiently informed to establish an order of society with a larger and more even distribution of benefit to all, and the law of will (force) is supplanted by the law of reason.

As defined in the Preamble to the Constitution, liberty is the absence of arbitrary human restraints upon personal conduct other than those imposed by the authority of just laws, obedience to which is an essential part of it.

Fundamental law. — The rights to life, liberty, and the pursuit of happiness are beyond the right of any government to legally usurp or infringe.

> To secure this (liberty) is the main business of governments and the reason for their institution. If they fall in this, they have failed in all. — *Blackstone.*

These principles were written by our fathers into a constitution of government, for the first time in human history, when they wrote the Constitution and it became the fundamental law of a new nation dedicated to the proposition that "all men are created equal" and that "government derives its just powers from the consent of the governed."

Equality. — What is meant by "equality" is clearly defined by Lincoln in his debate with Douglas.

> In responding to Douglas's question, "What do you mean, 'all men are created equal?'" Lincoln replied:

> I think the authors of that notable instrument intended to include all men, but they did not intend to declare all men equal in all respects. They did not mean to say all were equal in color, size, intellect, moral development, or social capacity. They defined with tolerable distinctness in what respect they did consider all men created equal — equal with "certain inalienable rights, among which are life, liberty, and the pursuit of happiness." This they said and this they meant. They did not mean to assert the obvious untruth that all were then actually enjoying that equality, nor yet that they were about to confer it immediately upon them. In fact, they had no power to confer such a boon. They simply meant to declare the right, so that enforcement of it might follow as fast as circumstances should permit.

107. Personal liberty. — *Freedom of action.* — Every citizen is on an equal footing as to privileges and opportunity. Any denial of such rights results from either the limited ability of the individual to take full advantage of opportunity, or because of prejudices in no way a part of the ruling law of our land.

Born free citizens, or acquiring that right through naturalization, we have full freedom of action — without infringement upon the rights of others — to reside or travel at home or abroad under the protection and with all privileges accorded by our Government, regardless of race, color, religion, or social station.

Full opportunity is here given to every citizen to work out his own ideals and ideas. To the native born, this privilege is accepted as a matter of no great significance, for he is wholly unfamiliar with the laws, traditions, and customs that direct and restrict individual action of citizens in foreign countries.

The American citizen frequently changes his occupation. His very liberty keeps him on the alert for an opportunity to better his financial or social status. The change is one of occupation, not of personality; his pride and self-respect are not involved.

108. Religious liberty. — No greater liberty was ever conferred on a people than that of freedom to worship according to the dictates of one's own conscience.

The first amendment to the Constitution declares that "Congress shall make no law respecting an establishment of religion, or prohibit the free exercise thereof."

All persons have the privilege to entertain any religious belief, practice any religious rite, teach any religious doctrine, which is not subversive of morality and does not interfere with the personal rights of others.

However, this liberty cannot be "invoked as a protection against legislation for the punishment of acts inimical to the peace, good order, and morals of society," because professed doctrines of religious belief are not superior to the laws of the land. No person is permitted to become a law unto himself, nor may he in the name of religion, or through a religious ceremony, violate the law.

> Religious liberty does not include the right to introduce and carry out every scheme or purpose which persons see fit to claim as a part of their religious system. While there is no legal authority to constrain belief, no one can lawfully stretch his own liberty of action so as to interfere with that of his neighbors, or violate peace or good order. — *United States Supreme Court.*

> Laws are made for the government of actions, and while they cannot interfere with mere religious beliefs and opinions, they may with practices. — *United States Supreme Court.*

Separation of church and state. — The separation of church and state is a fundamental principle of American Government. Neither is permitted to dictate to or exercise power over the other. In no other way can religious liberty be preserved.

Religion and national defense. — There is no place for the doctrine of "noncooperation." Religious beliefs will not excuse any citizen from rendering service in the defense of the country, although Congress has power at its discretion to exempt him.

109. Freedom of speech and press. — The right to act, to think, to speak, to print, is the surest way to protect the liberties, and continue the full measure of independence which America so richly possesses. In these rights lies the means of creating a public opinion representative of the entire Nation. This liberty is indispensable to further social, economic, and political development. Clash of opinions creates interest and thought on all public questions. A realization of the force of public opinion expressed by the ballot, awakens a sense of responsibility that compels the best minds to give careful study to any subject that vitally concerns our Nation. Through the present means of communication, the people are daily informed in every matter of national or international import.

Abuses. — This privilege does not permit the publication of libels or other matter injurious to public morals or private reputation. Like all liberties granted under the broad principles of the Constitution, these rights are abused to the detriment of the best interests of the people.

Propaganda. — Propaganda floods our country from every conceivable source. Active and vociferous agencies have been organized for the express purpose of advancing doctrines absolutely not in accord with the fixed principles of our Nation. In the most persistent manner, they seek to tear down rather than build, to destroy rather than improve. One of their most subtly dangerous features is that it is so camouflaged as to make it appear to have an innocent purpose.

To prevent such activities during the World War, Congress found it necessary to pass the espionage act of 1917 for the safety of the State and the successful outcome of the struggle.

We carefully supervise every agency whose business may, in any degree, affect the physical health of our people. Equal care should be exercised over all agencies which in any manner may affect our social, economic, or political life.

Restriction of abuses. — There is no law in any state or nation that prohibits freedom of speech or press, but there are laws against the abuse of this right. Restrictions may be necessary for the preservation of public order and the protection of the State. While Congress is forbidden by the Constitution to abridge the freedom of speech or the press, the punishment of those who violate every principle of loyalty and patriotism modifies in no manner the constitutional provision. The law punishes because of the crime against the country and its citizens.

> The first amendment "cannot have been, and obviously was not, intended to give immunity for every possible use of language." — **Justice Holmes.**
>
> **Blackstone's** maxims, which help to interpret the present limitation on speech and press:
>
> (1) Between public and private rights, the public rights must prevail.
>
> (2) Liberty to all, but preference to none.
>
> (3) Those offenses should be most severely punished which are most difficult to guard against.

110. Economic liberty. — *Property rights safeguarded.*
— Under no other government are property rights of the individual so provided with safeguards for their full protection. Property is at the base of civilization. Without incentive of right to its private possession and full protection against confiscation, no progress would be made in material betterment.

Economic liberty, the power of initiative, and the protection of property rights have developed a philosophy of life peculiar to America — the "dignity of work." Every American is expected to be a worker.

Based upon the constitutional assurance of the security of property, finance, and labor have joined in the creation of industry, making America the richest nation in the world. Her wealth has been distributed to the enrichment of her entire population.

111. Political liberty. — *Equal participation.* — The list of public office holders in city, State, and Nation reveals the measure of political liberty granted in America. There are found representatives of practically every race in the world. They have been elected by the people as their able and honorable representatives.

Every citizen enjoys the protection and benefits of our municipal, State, and National Governments.

Any suggestion of racial or religious differences is frowned upon. It is the sincere wish of the majority that tolerance and understanding weld our people of all nationalities into a social, economic, and political unity for the purpose of developing a strong national character and a race of men and women whose ideals and attainments shall be an inspiration and help to the peoples of all the earth.

The greatest degree of political liberty is secured by wise laws properly enforced. Anarchy destroys liberty because it is lawlessness and confusion, and utter disregard of all government.

112. Safeguards to our liberties. — By clinging to the ideas and ideals which animated the framers of the Declaration of Independence, we can assure not only peace within, but national security and respect from other nations.

When we fail to adequately comprehend the principles incident to our Government and its fundamental ideals which have made our Government, the United States of America faces anarchy and destruction.

> Let our object be our country, our whole country, and nothing but our country, and by the blessing of God may that country itself become a vast and splendid monument, not of oppression and terror, but of wisdom, of peace, and of liberty, upon which the world may gaze with admiration forever. — *Webster.*

QUESTIONNAIRE — LIBERTY
AND INDEPENDENCE

Describe the historical background of human liberty.
What foundation is necessary for liberty and independence?
What was the Declaration of Independence?
When are a people free and independent?
How do the "enemies within" show disrespect for the Declaration of Independence?
How, only, can the principles set forth in this document survive?
Name and describe the two kinds of liberty.
Does liberty mean freedom from control? Explain.
How is liberty defined in the Preamble to the Constitution? By whom and when?
Define personal liberty.
What is meant by religious liberty?
Do religious beliefs excuse a citizen from rendering service in defense of the country? Explain.
What is the relation of church and state?
Is freedom of speech and press beneficial to our national life?
What are some of its abuses? Describe.
Can these abuses be restricted? How?
State Blackstone's maxims which help interpret the present limitations on speech and press.
What safeguards are given to property?
What is meant by political liberty?
By what instrumentality can the greatest degree of political liberty be secured?

SECTION VIII

LESSON 8. — THE PURPOSE OF GOVERNMENT

113. **Progress of government.** — In the beginning of human history, with needs and wants limited to food and shelter, man's dominating impulse was the preservation of life.

His social instinct led to the establishment of families, groups, and tribes. Transmitting habits, traditions, customs, and superstitions to succeeding generations, there came to be formed definite and distinctive racial types with fixed governing principles.

Governments of a kind were set up, order was created, but with the accumulation of property, and increasing wants, conflicts occurred, the strong despoiling the weak. Alliances for defense and offense were formed.

Agreements between rulers and subjects and forms of treaties with nations brought about a more or less defined code of conduct and law, invariably enforced to the benefit of those who held the power.

Government by autocracy. — Selfish and often cruel leaders preyed upon the weak and ignorant in the accomplishment of their ambitious designs. Autocracy held power through appeal to the emotions engendered by pomp and glitter of the court or by fear created through control of military forces and the means of livelihood.

By various methods, the rights of citizenship were confined to the prescribed limits dictated by "will" (force) until increasing intelligence within the ranks of the people began to exert a counteracting influence.

The historical development of the "ancient liberties" of the English people, establishing individual rights, began with the meeting of King John and the Barons on the field of Runnymede in 1215 A. D., where the Magna Charta was signed, which guaranteed rights beyond the power of the king to take away. By successive steps, in protection of these rights, came the act of Parliament (1295), Petition of Rights (1628), the habeas corpus act (1679), Bill of Rights (1689), and the act of settlement (1701).

These liberties did not originate with charters but were simply confirmed by them and made the "fixed principles of freedom."

Restrictions of government on the life of the people created caste, favoritism and taxation became oppressive, and men left Europe and came to America.

Government of laws. — Until the adoption of the Constitution, government was imposed by the will of the minority and enforced by absolute control of economic institutions and military forces.

Under the Constitution, a "Government of laws and not of men" was formulated out of the experiences of the centuries in which feudalism, despotism, autocracy had given form to the ruling forces of government.

114. Sources of the Constitution. — The underlying principles of the Constitution were not formulated in a day. The three great American charters of liberty contained the fundamental principles of American government: "Bill for establishing religious freedom in Virginia," "Virginia Bill of Rights," and "Declaration of Independence." Before the Constitutional Convention met in Philadelphia, many plans and suggestions were drafted and presented to the convention.

> In addition to this careful preparation after more than a century of self-government, there were in the convention men of extraordinary natural ability and wide experience, like Washington, Franklin, and Hamilton. There were men who had studied law at the Inner Temple in London, who had been educated in the University of Edinburgh, who had been graduated from American colleges, who had been governors of States, chief justices of supreme courts, and men who had achieved distinction at the bar and in business life. Edmund Burke stated in the House of Commons in March, 1776, that more books of law were going to America than any other kind. Of the 55 members of the Constitutional Convention, 31 were lawyers. Blackstone's Commentaries were taught by Chancellor Wythe in William and Mary College before the Declaration of Independence. John Marshall, Thomas Jefferson, and James Monroe were among his pupils.
>
> When our Constitution was written, Harvard College (1636) had been sending out educated young men for just a century and a half, William and Mary College (1693) had been graduating learned youths for almost a century, Yale College (1701) had been contributing to the education of the people for more than three-quarters of a century, and Princeton (1746) had been teaching for half a century. The people were well prepared for their great endeavor. — *Thomas James Norton.*

115. The purpose of government. — A correct understanding of the purposes of government furnishes a remedy for erroneous and dangerous ideas threatening this country.

Government is instituted for the common benefit, maintaining order, and protecting life, liberty, and property.

> To secure liberty is the main business of governments and the reason for their institution. — *Blackstone.*

Paternalism. — The paternalism of communism, which provides both property and subsistence for the individual, is not a proper function of government. It results only in individual irresponsibility.

116.   The Preamble to the Constitution. — The Preamble to the Constitution is a most accurate and comprehensive statement of the purpose of government. It explicitly sets forth the fundamental purposes for which government is primarily organized. The brevity, simplicity, and directness of its original draft, after 150 years of experience, require no change.

> We, the people of the United States, in order to form a more perfect Union, establish Justice, ensure Domestic Tranquility, provide for the Common Defense, promote the General Welfare, and secure the Blessings of Liberty to ourselves and our Posterity, do ordain and establish this Constitution of the United States of America. — *Preamble to the Constitution.*

"We, the people." — The convention, which met in Philadelphia in 1787, adopted a Constitution based upon the proposition that a people are able to govern themselves.

Under the Articles of Confederation, the State assumed control. A single State might exercise veto power over the will of all the others.

In the government set up under the new Constitution, the power and rights of the people are the source and final authority. It derives its "just powers from the consent of the governed." For the first time in human history, "the people" assumed control and government became subject to their will.

Nowhere is American independence and self-reliance better exemplified than in the words, "We, the people."

> The people, the highest authority known to our system, from whom all our institutions spring and upon whom they depend, formed it. — *President Monroe.*

> Its language, "We, the people," is the institution of one great consolidated national government of the people of all the States, instead of a government by compact with the States for its agents. — *Patrick Henry.*

"A more perfect Union." — In the original federation, the States were but loosely joined. The Constitution was a demand for more effective control of the Union by the Government.

> In the efficacy and permanency of your Union, a government for the whole is indispensable * * *. You have improved upon your first essay (Articles of Confederation) by the adoption of a constitution of government * * * for the efficacious management of your common concerns. * * * Indignantly frown upon the first dawning of every attempt to alienate any portion of our country from the rest or to enfeeble the sacred ties which now link together the various parts. — *Washington — Farewell Address.*

In the course of the Civil War, the Southern States sought to dissolve our Union; President Lincoln sought to preserve our Union.

> The States have their status in the Union, and they have no other legal status * * *. The Union, and not themselves separately, procured their independence and liberty. * * * The Union is older than any of the States and, in fact, created them as States. — **Abraham Lincoln — Message to Congress, July 4, 1861.**

The right of secession was forever settled by the fourteenth amendment to the Constitution, which declares, "All persons born or naturalized in the United States and subject to the jurisdiction thereof are citizens of the United States and of the State wherein they reside." The National Government is not an assemblage of States, but of individuals.

To refuse allegiance to the United States is to be a traitor to the Nation. However, in the dual capacity of citizenship, we render service as citizens of both the State in which we hold legal residence and the United States. Each of our 48 States retains its own sovereignty in all matters relating exclusively to State affairs, in which it is protected by its own constitution. In all interstate, national, or international affairs, both the citizen and the State owe allegiance to the Union.

"Justice." — Our Government, assures "justice" in that it is a government of laws, not of men. In the heat of passion or sectional interest, in clashes between groups or questions of policy, no minority or bloc may enforce its will. Should a majority seek to injure the rights of an individual citizen, the power of veto resting in the President, or the power of the Supreme Court as an unbiased tribunal, will insist that justice be done.

A series of checks and balances, which prevent the selfish interests of either individuals or groups from exercising their will to the injustice of another, is provided by the Constitution.

> Wherever there is an interest and power to do wrong, wrong will generally be done, and not less readily by a powerful and interested party than by a powerful and interested prince. — *James Madison.*

> In questions of power, then, let no more be heard of confidence in man, but bind him down from mischief to the chains of the Constitution. — *Thomas Jefferson.*

"Domestic tranquility." — At the conclusion of the Revolutionary War, the Colonial States were bankrupt. Foreign credit was exhausted and could not be reestablished until a responsible central government was created. Soldiers remained unpaid long after the war was

ended. Colonies quarreled with each other over duties imposed upon the goods sold or bartered. Chaos and anarchy, disillusion, and despair prevailed, all because of a lack of proper organization and power in government.

The Government established under the Articles of Confederation "defrayed all expenses out of the common treasury" to which each State was supposed to contribute, but this was done in full only by New York and Pennsylvania. All non-enforceable obligations were left to conscience, individual or collective.

"Domestic tranquility" requires a measure of enforced responsibility, mutual faith, and harmonious and prosperous conditions. These are provided under the Constitution through the powers conferred upon the National Government regulating interstate affairs, making interchange of commodities, communication, transportation, and freedom of residence, occupation, and industry equal to all.

"Domestic tranquility" is further assured by religious freedom, free speech, and free press, thereby establishing interchange of thought which results in the creation of a national public opinion and brings within its influence every citizen, regardless of race, religion, financial condition, or social qualification.

"Common defense." — A country worth fighting for to establish was worth fighting for to preserve.

> The Congress shall have power to lay and collect Taxes, Duties, Imposts and Excises, to pay the debts and provide for the Common Defense and Welfare of the United States. * * * To declare War, grant letters of Marque and Reprisal, and make rules concerning captures on Land and Water; to raise and support Armies, but no appropriation of money to that use shall be for a longer term than two years. * * * To provide and maintain a Navy; To make rules for the Government and Regulation of the land and naval forces; To provide for calling forth the Militia; to execute the laws of the Union, to suppress Insurrections and repel Invasions; To provide for organizing,

arming and disciplining the Militia. — *Constitution, Article I, section 8.*

Attention is especially called to the limited period of two years as the length of time to be covered by any appropriation of money for the military forces. Without the consent of the people through their Representatives in Congress, any army created would fall to pieces for lack of funds. A great deal is said about the effort to "militarize" America through carrying out the provisions of the national defense act of 1920. This act was created by the people, for the people, to be paid for by the people. It can be killed by repeal or by refusal to make necessary appropriations. In the last analysis, the people are the military force of the United States; their employees, the Regular Army, National Guard, and Organized Reserves, are working for them, and in absolute obedience to rules and regulations laid down by their agent, the Congress.

> The United States is not solicitous, it never has been, about the methods or ways in which that protection shall be accomplished; whether by formal treaty stipulation or by formal convention, whether by the action of judicial tribunals or by that of military force. Protection, in fact, to American lives and property is the sole point upon which the United States is tenacious. — *William M. Evarts (1878).*

"General welfare." — The United States is a family of Commonwealths. Each State is possessed of its own natural resources, in the development of which it is necessary for its own best interests to have the full cooperation of every other State in exchange of raw materials, finished products, and farm produce. Its great land areas and mighty rivers are frequently the concern of several States or of the entire Nation.

It is within the power of Congress to appropriate funds for constructing canals, river and harbor development, and control irrigation projects where more than one State is interested, hard roads, and Postal Service; to regulate communications and transportation; and, through its various departments, perform such other services as will result in benefit to all citizens. This is not paternalism, but that protection of person and property which enables the citizen to obtain the greatest possible returns in the exercise of his own initiative.

"Blessings of liberty." — To secure the "blessings" of liberty was the fundamental purpose of the makers of the Constitution and its subsequent adoption. They include all the rights and privileges that a citizen of this country enjoys — a voice in the Government; freedom to worship according to the dictates of the individual conscience; freedom of speech and of the press; the lack of restriction upon all inherent individual rights.

The liberty of America is not that which permits the individual citizen to do as he pleases. He may so long as he does not interfere with the liberty of others. The liberty of the individual ends where the rights of others begin.

> We all declare for liberty, but in using the word we do not all mean the same thing. With some, the word liberty may mean for each man to do as he pleases with himself and the product of his labor; while with others the same word may mean for some men to do as they please with other men and the product of other men's labor. Here are two not only different but incompatible things called by the same name — liberty. And it follows that each of the things is, by the respective parties, called by two different and incompatible names — liberty and tyranny. The shepherd drives the wolf from the sheep's throat, for which the sheep thanks the shepherd as his liberator, while the wolf denounces him for the same act as the destroyer of liberty. — *Abraham Lincoln.*

The "blessings" which the citizen enjoys under our form of government are secured through "liberty under law," the enforcement of which is their only safeguard.

The purpose of our Government is to protect (not to provide) the property of its citizens; to guard his person (not to provide his subsistence) while he acquires the means of livelihood; to give every citizen equal opportunity in his chosen work and assure him of equal standing before the law.

Our Government is the most nearly perfect of all in securing individual rights and ensuring the blessings of liberty. In no other nation is equal opportunity and equal protection assured, with such equal division of reward for labor and services rendered.

117. The American philosophy of government. — The American philosophy of government emphasizes that —

(1) Individual rights are sacred and it is necessary to establish a government in the protection of these rights.

(2) All the powers of government are derived from the people, who retain the supreme authority over all delegated powers of government.

(3) Individual rights are not permitted to be exercised in the contravention of the rights of society. Individual liberty is always bounded by social obligations.

(4) Government is exercised for the purpose of protecting the individual in his rights.

(5) Governmental powers are delegated to the National, State, or local authority, and are limited in their exercise by provisions of the Constitution as interpreted and defined by the Supreme Court.

(6) All rights not thus delegated are recognized as the inviolable right of the individual citizen and cannot be usurped by any governmental power.

(7) The Government of the United States is not a democracy but a Republic.

James L. Tippins

QUESTIONNAIRE — THE PURPOSE
OF GOVERNMENT

At the time of the adoption of the Constitution what was the usual form of world government?

What was the principal distinction between "government of laws" and "government of men"?

What were the sources of the American Constitution?

What led the colonists to leave Europe and come to America?

Describe the doctrine of paternalism.

Is the paternalistic form of government efficient?

Define the true purpose of government.

Why is a correct understanding of the purposes of government necessary?

What is the Preamble to the Constitution? Quote it.

What is the source and final authority of government?

What is the meaning of "consent of the governed"?

How did the "Union" under the Constitution differ from that under the Articles of Confederation?

Does "dual capacity" of citizenship affect loyalty to the Nation?

How does the Constitution assure "justice" to the individual citizen?

How can "domestic tranquility" become possible in a nation composed of all races?

Who provides for the "common defense" of the Nation? How?

What is meant by "general welfare"?

What "blessings of liberty" are secured by our Constitution?

In general, what is the American philosophy of government?

SECTION IX

LESSON 9. — REPRESENTATIVE GOVERNMENT

118.   Early forms of government. — Until the eighteenth century, the world had little experience with republics. In the ancient world, Greece and Rome furnished early examples of attempts to form democratic governments. In Grecian cities, popular government was practiced, the free people directly making the laws. In Rome, the townsman passed laws to his own advantage. And in the so-called Venetian republic, the power was vested in a few nobles.

After the failure of many experiments in free government, the ancient world turned to monarchy, believing that the people were unfit to govern themselves. For centuries, political revolutions were struggles for better government, rather than self-government.

At the time of the Revolutionary War, the republican form of government was discredited throughout the world, monarchy and oligarchy being considered the proper forms of good government.

119.   Representative government. — *The American experiment.* — A few races qualified themselves for self-government. To establish that form of government was a long, hard struggle which culminated in the great American experiment.

The United States set up a distinct and different form of government, the product of distinct racial stocks and centuries spent in learning the principles and art of self-government. In practice, our form of government is the most nearly perfect in securing individual rights and ensuring the blessings of liberty.

It differs from previous forms in certain vital and fundamental principles which have come to be known as "American institutions." Among these is that of self-government by representation, which is "the golden mean between autocracy and democracy."

120. Comparative analysis. — The following comparative analysis shows the principal characteristics of the three forms of government:

Autocracy:
- Authority is derived through heredity.
- People have no choice in the selection of their rulers and no voice in making of the laws.
- Results in arbitrariness, tyranny, and oppression.
- Attitude toward property is feudalistic.
- Attitude toward law is that the will of the ruler shall control, regardless of reason or consequences.

Democracy:
- A government of the masses.
- Authority derived through mass meeting or any other form of "direct" expression.
- Results, in mobocracy.
- Attitude toward property is communistic — negating Property rights.
- Attitude toward law is that the will of the majority shall regulate, whether it be based upon deliberation or governed by passion, prejudice, and impulse, without restraint or regard to consequences.
- Results in demagogism, license, agitation, discontent, anarchy.

Republic:
- Authority is derived through the election by the people of public officials best fitted to represent them.
- Attitude toward property is respect for laws and individual rights and a sensible economic procedure.
- Attitude toward law is the administration of justice in accord with fixed principles and established evidence, with a strict regard to consequences.

Republic (cont.):
- A greater number of citizens and extent of territory may be brought within its compass.
- Avoids the dangerous extreme of either tyranny or mobocracy.
- Results in statesmanship, liberty, reason, justice, contentment, and progress.
- It is the "standard form" of government throughout the world.

A republic is a form of government under a constitution which provides for the election of (1) an executive and (2) a legislative body, who working together in a representative capacity, have all the power of appointment, all power of legislation, all power to raise revenue and appropriate expenditures, and are required to create (3) a judiciary to pass upon the justice and legality of their governmental acts and to recognize (4) certain inherent individual rights.

Take away any one or more of those four elements, and you are drifting into autocracy. Add one or more to those four elements, and you are drifting into democracy. — *Atwood.*

121. Superior to all others. — Autocracy declares the divine right of kings; its authority cannot be questioned; its powers are arbitrarily or unjustly administered.

Democracy is the "direct" rule of the people and has been repeatedly tried without success.

Our Constitutional fathers, familiar with the strength and weakness of both autocracy and democracy, with fixed principles definitely in mind, defined a representative republican form of government. They "made a very marked distinction between a republic and a democracy * * * and said repeatedly and emphatically that they had founded a republic."

Madison, in the Federalist, emphasized the fact that this government was a republic and not a democracy, the Constitution makers having considered both an autocracy and a democracy as undesirable forms of government while "a republic * * * promises the cure for which we are seeking."

> In a democracy, the people meet and exercise the government in person. In a republic, they assemble and administer it by their respective agents. — *Madison.*

> The advantage which a republic has over a democracy consists in the substitution of representatives whose enlightened views and virtuous sentiments render them superior to local prejudices and to schemes of injustice. — *Madison.*

The American form of government is the oldest republican form of government in the world, and is exercising a pronounced influence in modifying the governments of other nations. Our Constitution has been copied in whole or in part throughout the earth.

122. No direct action. — Under the representative form of government, there is no place for "direct action." The inherent characteristic of a republic is government by representation. The people are permitted to do only two things; they may vote once every four years for the executive and once in two years for members of the legislative body.

123. Methods of representative government. — Constitutional government may be set up under either a written or an unwritten Constitution.

An unwritten constitution. — An unwritten constitution consists largely of customs, precedents, conditions, and understandings, and is constantly changing; any party in power may enact legislation materially affecting the methods of government and the political rights of citizens.

A written constitution. — In the United States, the rights of the people are fully protected and the functions of government strictly defined in a written document — the Constitution. It is called a "rigid Constitution" because the legislative power has no authority to change it. It is subject to amendment only by the authority and action of the people through their representatives in Congress.

> The Congress, whenever two thirds of both houses shall deem it necessary, shall propose amendments to this Constitution, or, on the application of the legislatures of two-thirds of the several states, shall call a convention for proposing amendments, which in either case, shall be valid to all intents and purposes, as part of this Constitution, when ratified by the legislatures of three fourths of the several states, or by conventions in three fourths thereof, as one...mode of ratification may be proposed by the Congress; provided...no state, without its consent, shall be deprived of its equal suffrage in the Senate. — *Constitution, Article V.*

Since the adoption of the Constitution, our Nation has increased in population from 3,000,000 to more than 125,000,000 and has developed from a wilderness to the greatest industrial nation in the world. The adequacy of our Constitution is evidenced by the adoption of only 19 amendments to modify the principles set forth in the original document.

As a wall of protection, our written Constitution stands between the people and those who, through lust for power, or the temporary passions of the moment, or for any other reason, would trespass upon the rights of person or property.

124. Consent of the governed. — The original desire of the colonists was "only to have a voice" in the affairs of the Government.

> Governments are instituted among men, deriving their just powers from the consent of the governed * * *. We have petitioned for Redress in the most humble terms: Our

repeated Petitions have been answered only by repeated tyranny. — *Declaration of Independence.*

The situation so developed that the colonists totally dissolved "all political connection between them and the State of Great Britain," and established a new form of government based upon the "consent of the governed." "Consent" in the drafting and approval of the instrument of government and its subsequent amendment was a new feature.

125. "American Bill of Rights." — When the Constitutional Convention was drawing to a close several members who opposed the adoption of the Constitution suggested a number of amendments, which, they declared, "would make the Constitution acceptable to them."

While the Constitution already contained many provisions for the protection of the rights of the individual citizen, various States desired that it contain further written stipulations that would remove every possibility of doubt and prevent disputes by "leaving no matters to inference, implication, or construction."

It was contended that the provision of the suggested Bill of Rights contained "various exceptions not granted * * *. Why declare that things shall not be done which there is no power to do?"

> The tyranny of legislature is a most formidable dread at present, and will be for many years. That of the Executive will come in its time, but it will be at a remote period. — *Madison.*

Subsequently, many of these features were incorporated in the first 10 amendments, adopted in 1791 as supplements to the Constitution, and are called the "American Bill of Rights."

> The first 10 amendments embodied "guaranties and immunities which are inherited from our English ancestors."
> — *Supreme Court (1897).*

126. Enumeration of constitutional rights. — Individual rights formally guarded by original constitutional provisions:

No ex-post-facto laws.

No bill of attainder.

No suspension of privileges of habeas corpus.

Trial by jury and at places where the crimes were committed.

Definition of treason and limiting punishment.

Guaranty of immunity and privileges of all States to the citizens of each State.

No religious test before admission to public office.

To which the Bill of Rights added:

Right of peaceable assembly and petition to the Government for redress of grievances.

Freedom of religion, speech, and press.

Right of the people to keep and bear arms—militia.

Quartering of soldiers only as provided by law.

Protection against unreasonable searches.

Right of accused to indictment by grand jury with certain exceptions.

No compulsory testimony against self.

No deprivation of rights without due process of law.

No confiscation of private property for public use without just compensation.

Right of a speedy and public trial by an impartial jury.

Right to demand information concerning the nature and cause of accusation.

To be confronted with witnesses against him.

Compulsory process for obtaining witnesses in his favor.

Have assistance of counsel for defense.

Right of trial by jury in suits of common law where value and controversy shall exceed $20.

Protection of verdict of said jury.

No excessive bail required.

No imposition of excessive fines.

No infliction of cruel and unusual punishment.

Rights retained by the people shall not be denied nor disparaged.

Powers not delegated to the United States by the Constitution nor prohibited by it to the States are reserved to the States or to the people.

127. Government by representation. — The framers of the Constitution were opposed to direct government. The remedy sought was to be found in representative government. Madison declared that the object to which their efforts were to be directed was how to prevent a majority rule and to preserve the spirit and form of popular government. The representative form of government was their answer.

The United States shall guarantee to every State in the Union a republican form of government. — *Constitution, Article IV, section 4.*

Sovereignty was placed in the hands of the people. No authority was delegated to any department either of National or State Government except by the people through the provisions contained in the Constitution.

There could be no question but that by a republican form of government was intended a government in which not only would the people's representatives make laws and the agents administer them, but the people would also directly or indirectly choose the *Executive.* — *Cooley.*

128. Compromises. — In the establishment of our dual form of government, a spirit of compromise prevailed. The instrument offered by the makers of the Constitution

175

was the result of compromise, especially in regard to the matter of representation; the smaller States demanded equal representation with the larger. The compromise established two Houses of Congress: the Senate, in which each State was given equal representation; the House of Representatives, in which the membership was apportioned to the population. The functions of the two Houses of Congress were specifically stated and their powers definitely limited.

129. Separation of powers. — Members of the convention of 1787 feared the oppression of highly concentrated power, whether on the part of an individual or the ascendency of a parliamentary majority. Any suggested scheme to be satisfactory must limit the power of government rather than expand it.

Their plan of government provided for the division of power into three departments:

A legislative body working together in a representative capacity having power of appointment, power of legislation, power to raise revenues, power to appropriate funds for public expenditure.

An executive department whose duty was law enforcement and administration of the departments.

A judicial or law-interpreting department, at the head of which stands the Supreme Court.

130. Checks and balances. — These departments were separated from each other as far as possible, cooperating when necessary. Checks were placed upon each, preventing anyone from becoming absolute or despotic. They were likewise balanced against each other in such a manner as to preserve the equilibrium of government: States are balanced against the Central Government; House of Representatives is balanced against the Senate; Senate is balanced against the House of Representatives; executive authority is balanced by the legislative;

legislative department is balanced by the executive; judiciary is balanced against the legislative, executive, and State governments; Senate is balanced against the President in all appointment to offices and all treaties; people hold a balance against their own representatives through periodical elections.

Among the curbs and restrictions on the powers of the Central Government, the strongest checks are: Establishment of a smaller legislative body with less changing personnel and longer terms based on equality of representation, having coordinate legislative authority, with the exception of revenue bills, which originate in the House of Representatives, and treaties and appointments, which are committed to the President and the Senate; the public sentiment of an intelligent and conservative people; popular elections; short terms of office.

131. Federal judiciary. — To accomplish the uniform interpretation of the Constitution, a Federal court system was necessary, and it was provided that the judges should be appointed by the President, "with the advice and consent of the Senate."

Through the system of checks and balances, the safeguarding of the Constitution is charged to the Supreme Court. However, every judge in the land is also bound, under oath or affirmation, to support it and declare void any enactment which violates its provisions.

When a State court fails to fulfill this obligation, "its action is reviewable and reversible by the Supreme Court of the United States."

> This system which makes the judges the guardians of
> the Constitution provides the only safeguard which has
> hitherto been invented against unconstitutional
> legislation. — *Dicey.*

The courts keep each authority within its proper sphere, but they have the power to interfere only when a concrete case is brought before them for judicial consideration.

> One method of assault may be to effect in the form of
> the Constitution alterations which will impair the energy
> of the system and thus undermine what cannot be directly
> overthrown. — *Washington — Farewell Address.*

> A Constitution may be undermined by the passing of
> laws which, without nominally changing its provisions,
> violate its principles. — *Dicey.*

One of the exceptional features of our republican form of government is the independence of the Federal judiciary whose jurisdiction extends to all cases arising under the Constitution itself; cases arising under the Federal laws and treaties; cases affecting ambassadors, consuls, etc.; cases of admiralty and maritime jurisdiction; cases in which the United States is a party; controversies between States; cases commenced by a State against the citizens of another State; controversies between the citizens of the same State under land grants from different States; cases between American citizens and foreign states, citizens or subjects.

The balance of power has been preserved. The Constitution, as a whole, stands unshaken with but slight encroachments of one department upon the other.

132. Delegated national powers. — Under the plan set up under the Constitution, certain definite powers are delegated to the three departments of government.

Among the powers delegated to Congress are to —

Levy taxes.

Coin money.

Pay national debts.

Regulate commerce.

Establish uniform naturalization laws.

Establish the post office.

Provide for the common defense.

Declare war.

Raise and support armies.

Provide a navy.

Among the limitations placed on the powers of Congress are —

Apportionment of representation and direct taxes among the States is determined by population.

No money can be paid except by law.

All orders, resolutions, and bills must be sent to the President for his consideration.

Privilege of habeas corpus shall not be suspended except in case of rebellion or insurrection.

Among the powers delegated to the President are —

Execute the laws.

Commander in Chief of the Army and Navy.

Commission all officers of the United States.

Grant reprieves and pardons.

Make treaties by and with the advice and consent of the Senate.

Nominate judges of the Supreme Court.

Give information to Congress in formal messages.

Sign or veto orders, resolutions, and bills received from Congress.

133. Powers reserved to state and people — The President and Congress can exercise only those powers directly granted them by the Constitution. All powers not so delegated are reserved to the people.

> The enumeration of certain rights shall not be construed to deny or disparage others retained by the people. — *Amendments to Constitution, Article IX.*

> The powers not delegated to the United States by the Constitution nor prohibited by it to the States are reserved to the States, respectively, or to the people. — *Amendments to Constitution, Article X.*

134. Dangers to representative government. — Whenever the republican form of government has not achieved success, the difficulty has not been with the system but with its faulty application. Several dangerous experiments have been proposed, such as the initiative, referendum, recall, and the election of judges. Departures from constitutional principles threaten to impair the efficiency of our representative form of government, and if continued, will ultimately destroy it.

Centralization. — Originally, "every influence favored the supremacy of the State as the center of gravity in government." Conferring strong powers on the proposed central government was feared and avoided. With the development of industry, invention, business, and transportation, the different sections of the country were brought into such intimate and immediate contact that "the knell of State sovereignty was sounded and the supremacy of the Union became inevitable."

New and practical problems confront the Government, such as — increase of governmental business; rise of technical questions in government; popular demand for greater speed in Government action, and increased size and unwieldiness of legislative bodies.

Opposition to centralization of power in the National Government rests upon the general dislike of concentrated power, and its destructive influence on our philosophy of government.

Sectional and class legislation. — Nothing is more repugnant to the American citizen than special or class legislation. The founders of our Government sought unity rather than differentiation. The Civil War settled for all time the question of the indissolubility of the Union. The general welfare of the Nation forbids sectional or class legislation. There must be no preference to the North, East, South, or West. Our motto should be "America for all, and all for America."

Multiplicity of laws. — The modern tendency of government is to create innumerable laws as corrective or restrictive measures; appointment of special officers for their enforcement, with the consequent restriction of State, community, and personal rights, without regard to the fact that the majority is unprepared or not willing to accept or respond to the restrictions imposed. Relief from encroachment upon the rights of the people will come when each citizen better learns the art of self-government and exercises his right of franchise.

Socialism, communism, anarchy. — The problems of capital and labor, employer and employee, cannot be solved by unrepublican methods. The suggestion of special legislation is socialistic and communistic in its theory and wholly repugnant to the American character.

Socialism or communism, which negates property rights; anarchy which negates law; the substitution of "direct action" for representative government; a government ownership — all should be avoided as perils that threaten the very foundation of this Republic.

Ignorance of citizens. — Webster said, "On the diffusion of education among the people rests the preservation and perpetuity of our free institutions." In the early Colonies, one of the first buildings to be erected was the schoolhouse. Here was laid, developed, and subsequently spread the ideals of liberty. One of the foundation stones of representative government is education.

An intelligent and informed citizen is an asset to the Nation. The great educational system of America makes it possible for every citizen to best fit himself for the tasks of life. In the common schools, all are taught a common language, a knowledge of American traditions, ideals, and philosophy of government.

Through education, the barrier that separates the citizen from the greater enjoyment of his freedom is removed, a better understanding of American ideals is established, and the influence of subversive propaganda is, in large measure, destroyed.

135. Safeguards. — In order to assure perpetuity to our form of government, certain safeguards are necessary against encroachments both from within and without.

Direct responsibility to the people. — Having derived its "just powers from the consent of the governed," the Government of the United States is directly responsible to the people as the highest authority. The United States is governed by public opinion — by the ideas and feelings of the people at large. The frequency of elections and the short terms of office give the people control. By reason of this, our representatives are slow to attempt any official action overstepping the bounds of their authority or beyond the approval of their constituency.

Restricted immigration. — Immigrants who enter the United States to exploit her resources without a thought of contributing a share to the general welfare are a menace to our country. Many seeking a haven of relief from the oppressions of poverty, ignorance, and restrictions, a place where gain is made easy and burdens made light, come in the spirit of the belief that America owes them a good living, security, and peace, without a thought of the price that has been paid to obtain these blessings or the cost of their maintenance. Against these, America acclaims the fundamental right to close the door, for this is our home, and we have the right to select whom we will to enjoy its privileges and bounties.

America is basically made and refuses to any the right to alter the plans, destroy any part of the structure, or rebuild it to their liking.

Knowledge concerning the Constitution. — For a proper appreciation of our Government, the citizen should know what the Constitution is and what it contains.

> The selection and combination of these elements was a master achievement of vision, ability, and governmental genius on the part of the delegates to the convention. — *Atwood.*

He should thoroughly understand the purposes of government as set forth in the Preamble to the Constitution; that the Constitution established a strictly representative form of government; and the general provisions in regard to amending the Constitution, when "necessary." All of this is essential to his proper "regard for the sterling worth of our beneficent heritage."

The only antidote to the erroneous and dangerous ideas of government now rampant through the world and threatening America is a better understanding of the meaning, value, and importance of our American philosophy of government as set up in the Constitution.

This will most effectively meet the propaganda of communism in its attack on our social, economic, political, and national institutions, which aims to destroy the family as the foundation of society, our system of capitalism which has produced the great economic success of America, our republican form of government, and our spirit of patriotism.

> The preservation of the sacred fire of liberty and the destiny of the republican model of government are justly considered as deeply, perhaps as finally, staked on the experiment entrusted to the hands of the American people. — *Washington.*

James L. Tippins

If in our case, the representative system ultimately fail, popular governments must be pronounced impossible. No combination of circumstances more favorable to the experiment can ever be expected to occur. The last hopes of mankind, therefore, lest with us; and if it should be proclaimed that our example had become an argument against the experiment, the knell of popular liberty would be sounded throughout the earth. — *Webster.*

QUESTIONNAIRE
REPRESENTATIVE GOVERNMENT

Name three kinds of world governments.
What is an autocracy? Name the principal characteristics?
What is a democracy? Name the principal characteristics?
What is a republic? Name the principal characteristics?
Which form of government did the makers of the Constitution seek to establish?
Name the methods of representative government. Describe them.
What new feature of government was incorporated into the Constitution?
Describe the "American Bill of Rights."
What "individual rights" are formally guarded by the original Constitution?
What "rights" were added by the first ten amendments?
How was majority rule prevented and popular government preserved?
How were the differences as to representation compromised by the framers of the Constitution?
What is meant by "separation of powers"?
Explain "checks and balances."
Describe the Federal judiciary.
Enumerate the national powers delegated to Congress.
What limitations were placed on the powers of Congress?
Name and define several dangers to representative government.
Name and define the main safeguards of representative government.
How does restricted immigration benefit —
(1) The social life of America?
(2) The economic life of America?
(3) The political life of America?

SECTION X

LESSON 10. — PERSONAL RESPONSIBILITY

136. Responsibility of the present. — Civilization is built upon the experiences of the past. Any improvements that have been accomplished are the results of human achievement. No system of living has yet been devised that relieves the individual of his personal responsibility for the improvement of human society. By personal effort, each individual should pass on to posterity a civilization better than he found.

The sense of personal responsibility increases with the advancement of civilization. Not only have desires and wants multiplied, but with the advance of physical science, there is also a quickened moral sentiment and spirit of philanthropic sympathy; an increasing recognition of the responsibility of each man for his fellow citizen.

137. American civilization dynamic. — American civilization is expressed in "power," the power of the individual citizen in the driving force of his initiative, adventurous spirit, self-reliance and dogged energy. To the American, life is a great adventure.

Human wants, desires, ambitions, spur mankind to achievements. Never satisfied, ever progressing, civilization has constantly improved, and with the improvement have come burdens and complexities, which add more and more to the problems of human society.

Through equality of opportunity, America gives each individual citizen an equal chance, yet his ability, intelligence, and character distinguish and classify him as progress is made.

When America was new, she called upon the racial stocks of the world to give their best. Out of these, she has built a great nation.

The intelligent, though uneducated foreigner, might have continued to live in his native land without the slightest mental awakening. Once landing upon American soil, he quickly catches the spirit of his new environment, takes advantage of the free institutions, and finds opportunity for development to his fullest capacity.

In modern progress, America leads the world. The American citizen, whether native or foreign born, must recognize his obligations and assume his responsibilities not only to America but also to the entire world.

138. Individual responsibility. — In the very nature of the organization and form of our Government, our free institutions, and the lack of all authority and order other than that created by the dictum of the people, the security and perpetuity of America rests upon the individual responsibility of her citizens.

139. Education. — It is the duty of every citizen to obtain the best possible education. To shirk this responsibility is to be unworthy of the "blessings of liberty" and untrue to his own best interest. Every new device, discovery of science, enlarged market, added production, facility of communication and transportation, carries with it a demand for an educated citizenry. Society, economics, local and foreign politics, add their demands for educated leadership and participation. Greater opportunities await the educated and fewer the uneducated with each passing year. It is the responsibility of every citizen to become fully informed, for through education is found the only sure means of perpetuating and improving our social structure.

140. High standards. — Civilization is not a circle but a pyramid. At its base is found the constantly increasing mass of humanity. Out of this common material, the world has been busily engaged in building the structure of civilization.

No one is compelled to remain at the base of the pyramid who has within himself the ability to find his way up. From that base have come most of the great men in history. Few born in riches or high social position have ever achieved greatness.

By her system of Government, America is at the mercy of those at the base of the pyramid. If through individual initiative and proper leadership they win their way toward the apex, they lift America also. If they remain inert, ignorant, indifferent, they become the common prey of unscrupulous leaders who seek to weaken or destroy the structure of our Government.

National character. — National character is the sum of every citizen. The Nation has a right to expect each citizen to maintain high ideals, and he has a right to expect the same of his neighbor. The resulting measure of satisfaction should spur any right-thinking individual to such attainment. The actual worth of a citizen to himself, his community, his country, regardless of any other accomplishment, is based on the high quality and standard of his thinking. Obedience to higher impulses builds up self-respect without which no true success is possible.

Community and home. — The United States has been developed by a succession of communities, independent of each other, yet closely related in their social, economic, and political interests. The character of the community is determined by the character of its homes, and the character of its homes is determined by the character of the individual citizen. He is the only person upon whom responsibility for community and home can be placed.

141. Importance of active citizenship. — Good government is the particular responsibility of the individual citizen in whom final authority is vested. It will be no higher in its ideals nor just in its administration than the sum of our national character.

The first and paramount duty of every citizen is to have a firsthand knowledge of the Constitution of the United States. He should learn the accurate, comprehensive, and masterly statement of the six principles of government as contained in the Preamble, and the plan for setting up and maintaining our representative form of government. It is in this document that individual rights and fundamental duties are set forth.

American citizens are stockholders in a great corporation — the Government of the United States. They annually spend three and one-half billion dollars in the cost of government. One citizen out of 13 gainfully employed works for this corporation. Its operation requires understanding, supervision, and skillful management.

The citizen is the governor of this Republic through the exercise of his right to vote — the most sacred right of a free people. He selects its rulers and decides its issues. The proper exercise of this right requires honesty and intelligence, and a knowledge concerning the dangerous tendencies that are threatening our republican form of government. He should weigh the merits of both men and issues, feeling himself responsible for the selection of proper persons as the representatives to whom are entrusted the affairs of government.

Vote. — To preserve American institutions, a bigger and better vote is required — citizens must perform their political duties on election day.

The entire electorate must be taught not only to vote but to vote according to principle and informed opinion. Our institutions are endangered and are well worth saving. In the presidential years of 1920 and 1924, scarcely half of the voters of the country went to the polls. In 1926 only 33 per cent of the electorate participated. The ultimate result of such indifference upon a government based upon the principle of the majority is disastrous.

In 1928 more than 7,000,000 young citizens became eligible to vote for the first time. While the vote, and the whole vote, should be attracted to the polls, it must be remembered that an unintelligent vote safeguards nothing and is harmful in its effect.

Public service. — Many citizens are so engrossed in their personal affairs that they are not willing to devote sufficient time to the business of government, leaving most important matters to be decided by a minority.

The functions of citizenship are not confined to the enjoyment of personal rights — they also involve the protection of those rights. Unless the obligations of the individual citizen are fulfilled, our entire governmental structure, with all of its rights and privileges, is endangered. The indifference of individual citizens threatens the destruction of the "blessings of liberty."

Opportunity for patriotic service calls for leadership and ability, and too many citizens fail to respond to this obligation. Every citizen should assist in the administration of law and justice by willingness to render jury service—nothing is more imperative. He should bear a proportionate part of the burden of taxation without an attempt at evasion. He should respect the rights of others both by precept and example. He should be willing to assume the duties of any public office to which his fellow citizens may call him. He should be useful and loyal, aiding in all public undertakings through a whole hearted cooperation for the welfare of all.

In every national emergency, the people have produced their leader—George Washington, Abraham Lincoln. When diplomacy has failed, as in the World War, the people have "volunteered."

Law and order. — The best government is that in which justice is most evenly administered. The better our Government, the more prosperous and contented the people. Every time the citizen assists the administration of justice, he makes a material contribution to the welfare of all.

Every citizen should observe and respect the law. It is no excuse that if a certain law interferes with his personal habits, desires, or beliefs, he should disregard it. Absolute acquiescence in the decisions of the majority when legally expressed is the vital principle of republics.

It is your personal responsibility not to contribute to the defeat of justice either by evading the law or consenting to its evasion by others. Statutory laws are presumed to be just and for the benefit of all law-abiding citizens. No greater responsibility rests upon the citizen than to demand just laws and their enforcement. There is nothing more degrading, more destructive in its effect upon personal honor and character, than evasion of law, bribery of officers, or contributing to the delinquency of others.

Respect for its authority, compliance with its laws, acquiescence in its measures, are duties imposed by the fundamental maxims of true liberty. The basis of our political systems is the right of the people to make and alter their Constitution and Government. But the Constitution which at any time exists till changed by an explicit and authentic act of the whole people is sacredly obligatory upon all. The very idea of the power and the right of the people to establish government presupposes the duty of every individual to obey the established government. — *George Washington — Farewell Address.*

The highest test of good citizenship is obedience to all laws. We cannot develop and keep alive the high sense of civic duty and pride by half-hearted allegiance to the Constitution. There should be no such thing as an oath to support the Constitution with mental reservation. — **W. B. Swaney.**

The law of the State of Illinois provides that every male person above the age of 18 years must respond to the call of the police officer in securing and apprehending an offender, and provides a penalty for failure to do so. A good citizen will never hesitate to inform an officer of any criminal act of which he has knowledge and to assist in apprehending a criminal and aid the officer in his prosecution. Under the laws of Illinois, a person who has knowledge of a crime and conceals it is also a criminal.

142. Public opinion. — Within each community, there is an invisible government which we call "public opinion." Without this force, our courts and police would be powerless in their effort to control. Only in proportion as public opinion backs the law can it or will it be enforced. To protect the land from the overflow of our great rivers, we erect dikes along their banks. The moment a "sand boil" appears behind a dike, a crew is rushed to the place, and repairs are made to prevent a break that might bring disaster to thousands.

Public opinion, expressing the true character of home and community, is the dike that protects America from the overflow of crime, immorality, irreligion, and injustice, which, if allowed to break through, will do an irreparable damage to the free institutions of America.

Public opinion reaches an uncommonly high level because every citizen is called upon to express his own judgment in community and national affairs, and to work for the betterment of his town, county, State, and country.

It is your personal responsibility to mold and control public opinion.

143.   Responsibility cannot be transferred. — "Responsibilities gravitate to the man who can shoulder them and power flows to the man who knows how." The recognition of the inequality of ability and the equality of moral obligation makes individual responsibility distasteful to the defective citizen. Efforts are being made to supplant the individual responsibility of American citizens with "State responsibility," which destroys self-respect, ambition, and national character. It demands "State control," which not only promises to relieve the citizen of his individual responsibility, but it also deprives the individual citizen of his personal liberties.

It is the duty of every American citizen to prevent the destruction of our Republic and individualistic form of government by any such destructive political philosophy.

144. Our example of individual responsibility. — The closing words of the Declaration of Independence reveal the seriousness with which the signers fulfilled their personal responsibility:

> For the support of this Declaration with a firm reliance on the protection of Divine Providence, we mutually pledge to each other our lives, our fortunes, and our sacred honor.

QUESTIONNAIRE
PERSONAL RESPONSIBILITY

In what manner is the American civilization dynamic?

Upon what does the security and perpetuity of America rest?

In what way does education affect the responsibility of the American citizen?

Is the Government of America at the mercy of the people? Explain.

Upon what is national character based?

What determines the character of the community and the home?

What is the first obligation of an American citizen?

Name several political responsibilities that test upon every citizen.

What does the successful operation of our Government require?

What two things are necessary for the preservation of our American institutions?

To what degree have our citizens availed themselves of the right to vote?

Are the functions of citizenship confined to the enjoyment of personal rights? Explain.

How has personal responsibility in times of national emergency been met?

Does personal responsibility require respect for and obedience to all of the provisions of the Constitution?

What should be the attitude of the individual citizen in reference to the observance of the law? Of Federal

laws? Of State laws? Of municipal ordinances?

Why is public opinion of such a high standard?

Can individual responsibility be transferred? Explain.

What would be the effect of "State responsibility"?

How seriously did the signers of the Declaration of Independence assume their personal responsibilities?

SECTION XI

LESSON 11. — SELF-PRESERVATION

145. Self-preservation the first law of nature. — Possessed at first with a slight intelligence, man's reliance was upon his physical powers; though brutal in quality, they were necessary for the preservation of life.

By the successive steps of groups, tribes, and small states, mankind evolved better means of protection: cultivated intelligence; developed habits, customs, and laws, which in a measure abridged the need of physical force.

146. Preservation of life and property. — To ensure the preservation of life and property, America has written into her Constitution absolute guaranties. In no other country is life and property so hedged about with protective laws — all securing the inalienable rights of the individual citizen.

The preservation of these rights is a dominant principle of the American philosophy of government. It limits that government, in writing, to certain definite powers, and the right is reserved to discharge any and all governmental servants who infringe upon the written will of the people.

By the system of government set up by our Constitution, the people have been able to regulate the agencies of government and control and direct corporations, capital, and labor. Mighty as is their power, they must not infringe upon the rights of any private citizen. Neither must the individual citizen infringe upon the rights of another.

Self-preservation for every citizen is guaranteed by the Constitution and guarded by the Supreme Court of the United States

147. National defense the bulwark of self-preservation. — That which preserves our rights has the right to be preserved. The Declaration of Independence was a "scrap of paper" until made immortal by the blood and sacrifice of our patriotic ancestors. The sufferings of Valley Forge, the courage of Washington, the victory of Yorktown, secured American liberties and wrote this great document into the hearts of liberty-loving people.

> This colony (Massachusetts) is ready, at all times, to spend and be spent in the cause of America. — *Warren — Message to Continental Congress.*

When the Constitution of the United States was adopted, with the exception of a small area along the Atlantic coast, America was a wilderness. She had a population of approximately 3,000,000 people.

By the liberties granted and with unrestricted opportunity, the colonials and pioneers conquered the wilderness, converting it into a land of fertile fields, great industries, and contented homes, an achievement of little more than 100 years.

Freedom not a gift. — Freedom is not a gift. It has been bought and paid for in the sacrifices of peace and war. It is laid in long hours of toil, the swing of the ax in the forest, the campfire of the lonely pioneer, the sod house of the early settler, the community stockade and the frontier Army post. Freedom has traveled a long, hard road. None but the strong and courageous have possessed it, and by none others can it be retained.

148. **Preservation of philosophy of government.** — Some interpret American liberty as the opportunity to exploit the Nation's resources and people by propaganda that aims to destroy American institutions. Under the guise of freedom of speech and press, every possible effort is being made to undermine and destroy the blessings of liberty. The problem of national defense deals not only with the question of elements but it is also the question of the preservation of that philosophy of government devised by the founders of this Republic.

149. Preparedness a necessity. — With our growth of population, wealth, and standing among the nations, we have learned that lack of adequate preparation in time of peace was the most certain way to encourage attack by other nations.

The security of the Nation has been endangered and lives unnecessarily sacrificed because of insufficient training and an inadequate number of trained officers and soldiers to give instruction or assume command.

Wars have been begun, which would never have been declared had America been prepared. Wars have been prolonged through lack of material and trained men to carry them rapidly forward to a successful issue. Hardships have been suffered by lack of supplies.

Our lack of preparedness, with its rush of preparation, personnel inadequately trained, lack of matériel or its means of manufacture, plus the immediate danger to national existence, not only created all the elements required for hasty and extravagant expenditures of money, but caused the criminal sacrifice of many of our best American citizens.

The Preamble to the Constitution states that one reason for its establishment is "to provide for the common defense," assigning that duty to the Federal Government. The "people," through their representatives in Congress, declare war; the task of carrying on the struggle devolves on the Army and Navy.

A million men springing to arms overnight would evidence patriotism; but an army of a million untrained patriots in this advanced day of scientific warfare means annihilation.

150. America not militaristic. — Our Government, from its inception, has opposed the idea of militarism. So determined were the colonials to prevent any possible military dominance, they placed a positive check upon such control by making the constitutional provision that money for maintaining the Military Establishment could not be appropriated for a period longer than two years, thereby placing in the hands of each succeeding Congress the power to control through holding the purse strings of the Nation.

> **Copy Editor note:** Ironically, this led to the withdrawal of this manual!

Military training is not militaristic. On the contrary, it is greatly beneficial to the youth of America. It builds men physically, morally, and intellectually and inculcates obedience, self-control, leadership, and loyalty.

151. America not imperialistic. — The United States has acquired a clear title to every square inch of land which has been added to that of the original thirteen Colonies. All territory annexed to the United States since 1803 has been acquired either by treaty or purchase, except Texas and Hawaii, which were admitted to the Union by their own request. In the latter instance, however, $200,000 was paid as a compensation to Liliuokalani.

152. Destructive idealism. — The attempt to undermine the Nation from within is more serious than the threat of armed force from without.

An impractical and destructive idealism called internationalism is being propagated by certain foreign agitators and is being echoed and reechoed by many of the Nation's "intellectuals." Its efforts are to combat the spirit of patriotism, to destroy that spirit of nationalism without which no people can long endure. History teaches that in proportion as nations lose their sense of nationalism, they become decadent. Having lost their sense of pride in the traditions of the past, their respect for national standards, their love for country, their spirit of patriotism — the end is near.

Pacifism creates a spirit of compromise with the very factors which operate to weaken the American Government. It attempts to force the Government into poses of internationalism and false altruism, destructive of the real interests of the American people.

Pacifism is baneful in its influence. It promotes distrust of country; debases the spirit of nationalism; is destructive of patriotism; undermines the policy of national defense; cooperates with destructive forces for the overthrow of national ideals and institutions.

> Experience has taught us that neither the pacific dispositions of the American people nor the pacific character of their political institutions can altogether exempt them from that strife which appears beyond the ordinary lot of nations to be incident to the actual pride of

the world, and the same faithful monitor demonstrates that a certain degree of preparation for war is not only indispensable to avert disasters in the onset, but affords also the best security for the continuance of peace. — *Madison.*

153. Prepared leadership. — Leadership is as difficult to develop in the Army as in business. The methods that ensure success in one are applicable to the other. One of the aims of military training is to produce leaders. The more competent they become, the higher the position they are sure to attain. So efficient is the training received by the officers in the Regular Army that many are invited to resign and accept positions of grave responsibility in the business world. In comparative measure, efficiency in leadership is also developed in enlisted men, in students of the Reserve Officers' Training Corps, and in trainees of the citizens' military training camps.

Business invariably gives preference to the young man who has had training in military leadership. Many industries provide their employees with 30 days' vacation on pay for the purpose of attendance at a summer training camp, knowing that they will return to their employment better equipped, better disciplined, and in every way much more valuable to themselves and their employers.

All the wars of the future will include science and machinery. Trained men will be needed to efficiently use these materials, for efficient leadership, education, skill, technique, training, and thorough discipline are as necessary as loyalty and willingness to serve.

154. Military policy of the United States. — The military policy of the United States is defensive, not offensive. America will go to war only in defense of the Nation, and no other nation need maintain a ship or a soldier as protection against a war of aggression instituted by the United States. America desires no territory belonging to other peoples. She seeks only self-preservation and the privilege of self-determination in peace with all the nations of the earth.

> Safety from external danger is the most powerful dictation of national conduct. — *Hamilton.*

> The genius and character of our institutions are peaceful * * * and the power to declare war was not conferred upon Congress for the purposes of aggression or aggrandizement, but to enable the General Government to vindicate by arms, if it should become necessary, its own rights and the rights of its citizens. — *United States Supreme Court.*

155.   The State Department. — By the means of arbitration and treaties, the State Department endeavors to settle international disputes. It is only after such methods have failed that the United States enters into war to enforce or protect its principles.

America has always endeavored to maintain peaceful relations with other nations. Yet practically every generation has been compelled to take up arms either in defense of the Nation or the principles set forth in her Constitution.

The attitude of the American Government toward other nations is —

> To cherish peace and free discussions with all nations having corresponding dispositions; to maintain sincere neutrality toward belligerent nations; to prefer in all cases amicable discussion and reasonable accommodation of differences to a decision of them by an appeal to arms; to exclude foreign intrigues and foreign partialities, so degrading to all countries and so baneful to free ones. — *Madison.*

156. National defense act. — The national defense act of 1920, amended to include March 4, 1927, provides:

> That the Army of the United States shall consist of the Regular Army, the National Guard while in the service of the United States, and the Organized Reserves, including

the Officers' Reserve Corps and the Enlisted Reserve Corps.

Except in time of war or similar emergency when the public safety demands it, the number of enlisted men in the Regular Army shall not exceed 280,000, including the Philippine Scouts.

The total authorized number of enlisted men, not including the Philippine Scouts, is at present fixed at 125,000.

Regular Army. — The Regular Army consists of approximately 118,000 enlisted men and some 11,500 officers. A large part of this force is used for garrison purposes at home and abroad. Those at home spend about eight months of the year in their own training and in intensive preparation for the work required of them in summer training camps.

The Regular Army also conducts the training of the Reserve Officers' Training Corps, the Organized Reserves, and the National Guard. Officers and men of the Regular Army are qualified to impart physical, mental, and moral training of the highest character. The very nature of their work makes them specialists in this field. No business or profession demands stronger character and ability. No group is more carefully disciplined, and nowhere will be found greater loyalty and honor. To train with and serve under the officers and enlisted men of the Regular Army is to be afforded an opportunity for personal betterment, which any wide-awake young American should be eager to accept.

National Guard. — The second amendment to the Constitution provides that —

A well-regulated Militia being necessary to the security of a free State, the right of people to keep and bear Arms shall not be infringed.

Prior to the national defense act of 1916, it was left to the States to provide an organized militia adequate in numbers, equipment and training to police the State in time of riot or insurrection; it was also to be used by the National Government in time of war with a foreign power. With the addition of a small standing Army, the forces thus provided were presumed sufficient for national defense.

Under the national defense act of 1920, the National Guard, in time of peace, is under the command of officers appointed by the governor of the State, but their training and administration is supervised by officers of the Regular Army assigned for that purpose. In time of war, the National Guard, as a component of the Army of the United States, is immediately called into national service. Together with the Regular Army, it serves as the first line of defense while the reserve forces are being organized and equipped.

> An efficient Militia is authorized and contemplated by the Constitution and required by the spirit and safety of free government. — *Madison.*

Organized Reserves. — The Organized Reserves, together with the other components of the Army, form the basis for a complete and immediate mobilization for national defense in any national emergency declared by Congress. Each reserve unit is now organized with its officers and a few enlisted specialists. In time of war, these units will assemble at points designated, there to be equipped and trained. Every member of the Reserve Officers' Training Corps and all graduates of the citizens' military training camps who have qualified for leadership and have been commissioned would be required to report to his proper station on the designated day.

> To expose some men to the perils of the battle field while others are left to reap large gains from the distress of their country is not in harmony with our ideal of equality. — *President Coolidge.*

157. Preparedness an agency for peace. — The desire for peace is the spirit of America, but that peace must be dynamic, not a peace characterized by weakness of purpose or lack of courage.

"Common defenselessness" is in opposition to the spirit of the Constitution. The best guaranty of peace is a physically fit people inspired by the spirit of the Constitution and strong enough to defend themselves against any foe.

True Americans should be prepared to defend our Nation against those influences that will not only destroy all patriotic ideals that have been acquired through years of struggle but which advocate the overthrow of our Government by force. Our very freedom allows enemies within to operate with appalling boldness. They have powerful allies in the persons of those who would abolish all of our defenses — who would have peace at any price.

The writings and utterances of the men who laid the foundations upon which posterity has been called to erect the superstructure of this Nation continually remind the citizen of the necessity to provide for an adequate defense of the blessings of liberty that, to insure them for future generations, we must be strong enough to protect and defend our country and our institutions from any hostile aggression, whether from without or within.

> By diffusing through the mass of the Nation the elements of military discipline and instruction; by augmenting and distributing warlike preparations applicable to future use; by evincing the zeal and valor with which they will be employed, and the cheerfulness with which every necessary burden will be borne, a greater respect for our rights and a longer duration of our future peace are promised than could be expected without these proofs of the national character and resources.
> — *Madison.*

158. Moral qualities essential to self-preservation. — The American citizen must emphasize those qualities of character which mark him as truly worthy of the privileges of independence and liberty. His claim to self-respect is sound only as he upholds the self-respect of his fellow citizens. His honor is sacred only as he protects the honor of his country. He values liberty and independence only in so far as he is willing to pay the price for its protection.

It takes more than eloquent speeches and hot words to accomplish sublime purpose — it takes risk; it takes sacrifice. It takes the spirit of a Nathan Hale, who, having been sent by General Washington to bring intelligence concerning the British in New York City, was captured within the British lines and executed as a spy by order of Sir William Howe, the British commander. His last words were: "I only regret that I have but one life to lose for my country." This is the spirit that won our liberties. It takes the same spirit to preserve our liberties.

> We mutually pledge our lives, our fortunes, and our sacred honor. — *Signers of the Declaration of Independence.*

The moral qualities essential to self-preservation are —
The will to win.
The courage to endure.
The willingness to die.

QUESTIONNAIRE
SELF-PRESERVATION

How is the preservation of life and property assured?

What is the bulwark of self-preservation? Explain.

How can the American philosophy of government be preserved?

Why is preparedness necessary?

Who declares war?

Is America militaristic? Explain.

How does military training benefit the youth of America?

Is America imperialistic? Explain.

What is meant by "internationalism"?

What are some of the baneful influences of pacifism?

What are the essential qualifications of leadership?

How can we best provide for the peace and security of our Nation?

Describe the military policy of the United States.

How does the State Department contribute to peace?

What is the national defense act?

What are some of its provisions?

Name and describe the three components of the Army of the United States.

In what way is preparedness an agency for peace?

What moral qualities are essential to self-preservation?

What provisions for national defense are contained in the Constitution?

Why should military service in time of war be determined by the National Government instead of the State or the individual?

SECTION XII

LESSON 12. — THE AMERICAN FLAG

> When flown with other flags.
> International usage.
> General uses.
> Reveille and retreat.
> Memorial Day.
> Unveiling statues.
> Military funerals.
> Patriotic occasions.
> Signal of distress.

159. Design accepted. — General George Washington, Robert Morris, and Colonel George Ross were appointed as a committee by the Continental Congress to produce a flag for the United States of North America. Their report was approved and the design adopted on the 14th of June, 1777. By resolution, Congress decided that the flag of the 13 United States should be 13 stripes, alternate red and white, and that the Union be 13 white stars on a blue field.

160. Significance of elements. — In describing its design, Washington said: "We take the stars from heaven, the red from our mother country, separating it by white stripes, thus showing that we have separated from her, and the white stripes shall go down to posterity representing liberty."

The Continental Congress defined the special significance of the chosen colors to be: White, suggesting purity and innocence; red, hardness and valor; blue, vigilance, perseverance, and justice.

The stars of the Union were not merely a collection but a new constellation representing a new ideal in political and governmental affairs. The newly formed States were to develop under the control of laws, not independently nor indifferent to each other — but a Union, one and inseparable.

161. Progress of the flag. — After 1812, the flag moved west with the pioneers who explored the vast regions beyond the Alleghenies, the Mississippi Valley, the Rocky Mountains, onward to the shores of the Pacific Ocean, and the islands of the sea. Representing the United States, the flag flies to-day in Alaska, Hawaii, the Philippines, Porto Rico, Guam, Tutuila, Panama, and at the North Pole.

> To be born under the American flag is to be the child of a king, and to build a home under the Stars and Stripes is to establish a royal house. Alone of all flags, it expresses the sovereignty of the people, which endures when all else passes away. Speaking with their voice, it has the sanctity of revelation. He who lives under it and is loyal to it is loyal to truth and justice everywhere. He who lives under it and is disloyal to it is a traitor to the human race everywhere. What could be saved if the flag of the American Nation were to perish? — *President Coolidge.*

162. Allocation of the stars. — President William H. Taft on October 25, 1912, by Executive order, designated the specific location of the stars and their definite representations. They were to be arranged in six rows of eight stars, each star to symbolize a State in the order of its ratification of the Constitution:

1. Delaware.	25. Arkansas.
2. Pennsylvania.	26. Michigan.
3. New Jersey.	27. Florida.
4. Georgia.	28. Texas.
5. Connecticut.	29. Iowa.
6. Massachusetts.	30. Wisconsin.
7. Maryland.	31. California.
8. South Carolina.	32. Minnesota.
9. New Hampshire.	33. Oregon.
10. Virginia.	34. Kansas.
11. New York.	35. West Virginia.
12. North Carolina.	36. Nevada.
13. Rhode Island.	37. Nebraska.
14. Vermont.	38. Colorado.
15. Kentucky.	39. North Dakota.
16. Tennessee.	40. South Dakota.
17. Ohio.	41. Montana.
18. Louisiana.	42. Washington.
19. Indiana.	43. Idaho.
20. Mississippi.	44. Wyoming.
21. Illinois.	45. Utah.
22. Alabama.	46. Oklahoma.
23. Maine	47. New Mexico.
24. Missouri.	48. Arizona.

163. Inspiration of the flag. —Like the cross, the flag is sacred. It represents the living country and is itself considered a living thing. It flies not only as the symbol of organization and protection; it also calls to duty. To the flag of the United States, and all that it represents, every citizen of America should render respect, reverence, and devotion.

As you feel about your flag, so you feel about your Nation.

Your flag, my flag, our flag! May we honor her as she honors us!

164. The future of the flag. —This flag, the emblem of justice and government, stands for the just use of undisputed national power. No nation is going to doubt our power to assert its rights.

> It is henceforth to stand for self-possession, dignity, for the assertion of the right of one nation to serve the other nations of the world — an emblem that will not condescend to be used for purposes of aggression and self-aggrandizement; that it is too great to be debased by selfishness; that has vindicated its right to be honored by all nations of the world and feared by none who do righteousness. — *Woodrow Wilson.*

165. Kinds of national flags. —There are four kinds of national flags: Flags which are flown at military posts or on ships and used for display generally; small flags or ensigns which are used on small boats; colors which are carried by unmounted regiments; and standards which are carried by mounted regiments, and are, therefore, smaller in size than colors.

There is prescribed in Army Regulations a knotted fringe of yellow silk on the national standards of mounted regiments and on the national colors of unmounted regiments. However, there is no law which either requires or prohibits the placing of a fringe on the flag of the United States. Ancient custom sanctions the use of fringe on the regimental colors and standards, but there seems to be no good reason or precedent for its use on other flags.

166. Federal laws. — There is no Federal law now in force pertaining to the manner of displaying, hanging, or saluting the United States flag, or prescribing any ceremonies that should be observed in connection therewith.

There are but four Federal laws on the statute books that have any bearing upon this subject:

(1) The act of Congress approved February 20, 1905, providing that a trade-mark cannot be registered which consists of or comprises "the flag, coat of arms, or other insignia of the United States, or any simulation thereof."

(2) A joint resolution of Congress approved May 8, 1914, authorizing the display of the flag on Mother's Day.

(3) The act of Congress approved February 8, 1917, providing certain penalties for the desecration, mutilation, or improper use of the flag within the District of Columbia.

(4) The act of Congress approved May 16, 1918, providing, when the United States is at war, for the dismissal from the service of any employee or official of the United States Government who criticizes in an abusive or violent manner the flag of the United States.

Several States of the Union have enacted laws which have more or less bearing upon the general subject, and it seems probable that many counties and municipalities have also passed ordinances concerning this matter to govern action within their own jurisdiction.

> No present Federal statute (exists) punishing the desecration or abuse of the flag, in time of peace or in time of war. — *Attorney General John G. Sargent.*

> A majority of States have passed acts designed to punish the desecration of the National flag and to prevent its use for advertising purposes. The constitutionality of such State legislation was upheld by the Supreme Court in Halter v. Nebraska, 205 U. S. 34.

> There is a Federal statute similar in terms to many of the State laws which punishes the improper use of the flag in the District of Columbia — act February 8, 1917, chapter 34 (39 Stat. 900), but there is no Federal enactment which punishes such use outside the District.

167. Method of displaying the flag. — There are certain fundamental rules of heraldry which indicate the proper method of displaying the flag. There are also certain rules of good taste, which, if observed, would ensure the proper use of the flag.

(1) The union of the flag is the honor point; the right arm is the sword arm and, therefore, the point of danger and hence the place of honor.

(2) When the national flag is carried, as in a procession, with another flag or flags, the place of the national flag is on the right — i. e., the flag's own right.

(3) When the national flag and another flag are displayed together, as against a wall from crossed staffs, the national flag should be on the right, the flag's own right — i. e., the observer's left — and its staff should be in front of the staff of the other flag.

(4) When a number of flags are grouped and displayed from staffs, the national flag should be in the center or at the highest point of the group.

(5) When the national flag is hung either horizontally or vertically against a wall, the union should be uppermost and to the flag's own right — i. e., to the observer's left. When displayed from a staff projecting horizontally or at an angle from a window sill or the front of a building, the same rules should be observed.

(6) When the flag is suspended between buildings so as to hang over the middle of the street, a simple rule is to hang the union to the north in an east and west street or to the east in a north and south street.

When flown with other flags. — When flags of States or cities or pennants of societies are flown on the same halyard with the national flag, the national flag must always be at the peak. When flown from adjacent staffs, the national flag should be hoisted first. There is a chaplain's flag authorized in Army Regulations, but there is no church pennant prescribed. Neither the chaplain's flag nor any other flag or pennant is authorized to be placed above or to the right of the national flag.

International usage. — The display of the flag of one nation above that of any other nation in time of peace is forbidden. When the flags of two or more nations are to be displayed, they should be flown from separate staffs or from separate halyards, of equal size and on the same level.

General uses. — There is no Federal law governing the subject, but it is suggested —

That the national flag, when not flown from a staff, be always hung flat, whether indoors or out.

It should not be festooned over doorways or arches nor tied in a bowknot nor fashioned into a rosette.

When used on a rostrum, it should be displayed above and behind the speaker's desk. It should never be used to cover the speaker's desk nor to drape over the front of the platform. For this purpose, as well as for decoration in general, bunting of the national colors should be used, arranged with the blue above, the white in the middle, and the red below.

Under no circumstances should the flag be draped over chairs or benches, nor should any object or emblem of any kind be placed above or upon it, nor should it be hung where it can be easily contaminated or soiled.

No lettering of any kind should ever be placed upon the flag. It should not be used as a portion of a woman's costume nor of a man's athletic clothing. A very common misuse of the flag is the practice of embroidering the flag on cushions and handkerchiefs, and the printing of the flag on paper napkins. These practices, while not strictly a violation of any present Federal law, certainly are lacking in respect and dignity and cannot be considered as evidence of good taste.

There is no objection to flying the flag at night on civilian property, provided it is not so flown for advertising purposes.

Reveille and retreat. — It is the practice in the Army, each day in the year, to hoist the flag briskly at sunrise, irrespective of the condition of the weather, and to lower it slowly and ceremoniously at sunset, indicating the commencement and cessation of the activities of the day.

Memorial Day. — On Memorial Day (May 30) at all Army posts and stations, the national flag is displayed at half-staff from sunrise until noon and at full-staff from noon until sunset.

When flown at half-staff, the flag is always first hoisted to the peak, the honor point, and then slowly lowered to the half-staff position in honor of those who gave their lives to their country, but before lowering the flag for the day, it is raised again to the head of the staff, for the Nation lives and the flag is the living symbol of the Nation.

Unveiling statues. — When flags are used in connection with the unveiling of a statue or monument, they should not be allowed to fall to the ground, but should be carried aloft to wave out, forming a distinctive feature during the remainder of the ceremony.

Military funerals. — When the national flag is used on a bier or casket at a military funeral, the rule is the reverse of that for hanging vertically against a wall. The union should be placed at the head of the casket and over the left shoulder of the soldier. The casket should be carried foot first. The flag should not be lowered into the grave, and in no case should it be allowed to touch the ground.

When a body is shipped to relatives by the War Department for private burial, the flag which drapes the shipping case is turned over to relatives with the remains for use at the funeral, and may be retained by them.

Patriotic occasions. — It is becoming the practice throughout the country among civilians to display the national flag on all patriotic occasions, especially on the following days: Lincoln's Birthday, February 12;

Washington's Birthday, February 22; Mother's Day, second Sunday in May; Memorial Day, May 30; Flag Day, June 14; Independence Day, July 4; and, Armistice Day, November 11.

In certain localities, other special days are observed in the same manner.

Signal of distress. — The flag should never be hung nor displayed union down except as a signal of distress at sea.

168. Disposition of worn-out flags. —Old or worn-out flags should not be used either for banners or for any secondary purpose. When a flag is in such a condition that it is no longer a fitting emblem for display, it should be destroyed, preferably by burning or by some other method lacking in any suggestion of irreverence or disrespect to the emblem representing our country.

169. Military salute to the flag. — When officers and enlisted men pass the national flag not incased or when the national flag is carried in a parade or procession, they will render honors as follows: If in civilian dress and covered, they will uncover, holding the headdress opposite the left shoulder with the right hand; if in uniform, covered, or uncovered, or in civilian dress uncovered, they will salute with the right-hand salute.

170. National anthem. ⊢ The musical composition familiarly known as the Star-Spangled Banner is designated as the national air of the United States of America. When played, all officers and enlisted men present and not in formation are required to stand at attention, facing the music, except when the flag is being lowered at sunset, on which occasion they are required to face toward the flag. If in uniform, they shall render the prescribed salute at the first note of the anthem, retaining the position of salute until the last note of the anthem. If not in uniform and covered, they are required to stand and uncover at the first note of the anthem, holding the headdress opposite the left shoulder until the last note is played, except in inclement weather when the headdress may be held slightly raised. The custom of rising and remaining standing and uncovered while the Star-Spangled Banner is being played has grown in favor among civilians.

The Star-Spangled Banner should be played through without repetition of any part not required to be repeated to make it complete. It should not be played as part of a medley nor for dance music, nor at any point in a program or performance except at the beginning or the end. It is the practice in the Army to play the Star-Spangled Banner at the end of a musical program.

171. National salute. — The national salute to the American flag requires one gun for every star.

NOTE. — It is not within the province of the War Department to force upon persons not in the military service the regulations governing the use of the flag within the Army.

172. Initial events of the American flag.

June 14, 1777: The first American flag, made by Betsy Ross, was adopted by the Continental Congress as the flag of the United States of North America.

1787-1790: The Stars and Stripes was first carried around the world by the ship *Columbia*.

August 2, 1777: An improvised Stars and Stripes was hoisted at Fort Stanwix, N. Y.

November 1, 1777: The American flag was first flown at sea by Captain Paul Jones. He sailed to carry the news to France that Burgoyne had surrendered.

February 14, 1778: The first salute given the American flag, at Quiberon Bay, France, when the French Admiral La Motte Piquet, saluted the flag on the *Ranger*, commanded by Captain Paul Jones.

September 11, 1777: The American flag first went into battle, receiving its baptism of blood at the Brandywine.

September 13, 1814: Francis Scott Key wrote the Star-Spangled Banner during the battle at Fort McHenry, in Baltimore Harbor. It was later officially designated as the national anthem.

July 24, 1866: The first American flag manufactured from American material was hoisted over the Capitol at Washington. Previously, the bunting had been manufactured outside the United States.

QUESTIONNAIRE — THE AMERICAN FLAG

Explain the significance of the "elements" in the American flag.
Describe the "progress" of the flag.
Which star is allotted to your State? Why?
What is meant by the "inspiration of the flag"?
What is its message to you?
There are how many kinds of national flags? Name them.
What Federal laws relate to the flag?
Should the American flag ever be used as an advertising device? Explain.
Describe the methods of displaying the flag.
When flown with other flags, what is the position of the American flag?
What is the international usage?
What suggestions are made as to general uses?
How is the flag flown on Memorial Day, and what is its significance?
In unveiling statues, how should the flag be used?
Describe the use of the flag at military funerals.
On what special occasions is it customary to display the |American flag?
What is the position of the flag when used as a signal of distress?
Describe the military salute to the flag.
How should the national anthem be played? What should the audience do?
What is the national salute to the flag? Explain.

CONSTITUTION OF THE UNITED STATES OF AMERICA

FOREWORD

Our Constitution is the foundation upon which this Republic rests. It is now the oldest written constitution functioning in the world and is quite generally conceded the wisest plan of government ever conceived.

Under its beneficent influence, we began to solve problems and secure individual comforts and privileges that had baffled philosophers and statesmen for ages. We have harmonized into a splendid and loyal citizenship people of many nationalities coming to our shores with varying ambitions and ideals, and have made orderly progress unparalleled in history until we have become the leading nation of the world.

In studying the Constitution, it is essential to have clearly in mind what portions have been modified or supplanted by amendment and what portions have become obsolete. The changes are clearly indicated in this edition.

When the Constitution was written, our country was in a condition of bankruptcy, chaos, and anarchy. Within three years after its adoption, a most favorable condition for orderly progress had been established. That beneficent transformation wrought by the Constitution is one of the most amazing facts in all history.

The men who wrote the Constitution had great mental acumen, political understanding, and moral courage. Their lives had been devoted largely to study and thought concerning government and to rendering public service. They were politically minded in the sense that Edison and Marconi are electrically minded; that Lindbergh and Chamberlain are aviation minded; that Socrates and Emerson were philosophically minded; that Newton and Kepler were scientifically minded.

To regard the Constitution merely as a statement of principles and an enumeration of rights and guarantees results in confusion and a false concept. It is a statement of the purposes of government, and the statement of a plan for setting up and administering a Federal representative government in harmony with the purposes to which it was dedicated.

Every proper activity of government can be classified under one or more of the six great purposes set forth in the Preamble. The plan for the division of powers into legislative, executive, and judicial departments, combining proper independence with means for helpful cooperation between those departments under well balanced restraints, makes possible a scientific administration of government.

The Constitution is very much the kind of a plan for handling the problems of government that the alphabet is for handling the problems of language; that the scale is for handling the problems of music; that the ten digits are for handling the problems of arithmetic.

Notwithstanding the vital importance of the Constitution to the well-being of this Republic, the number of persons who know much about it is tragically small. Increasing knowledge of its meaning and value will bring increasing desire for better understanding.

Harry Atwood

THE CONSTITUTION OF THE UNITED STATES

Edited by HARRY ATWOOD

Showing all portions of the original Constitution which have become obsolete enclosed in brackets in bold type, and all portions which have been modified or supplanted by amendment in bold italics, with notes indicating the amendments by which the changes were made

PREAMBLE

We, the people of the United States, in order to form a more perfect union, establish justice, insure domestic tranquility, provide for the common defense, promote the general welfare, and secure the blessings of liberty to ourselves and our posterity, do ordain and establish this Constitution for the United States of America.

ARTICLE I

THE LEGISLATIVE DEPARTMENT

Section 1. All legislative powers herein granted shall be vested in a Congress of the United States, which shall consist of a Senate and House of Representatives.

THE HOUSE OF REPRESENTATIVES

Sec. 2. (1) The House of Representatives shall be composed of members chosen every second year by the people of the several States, and the electors in each State shall have the qualifications requisite for electors of the most numerous branch of the State legislature.

(2) No person shall be a Representative who shall not have attained to the age of twenty-five years, and been seven years a citizen of the United States, and who shall not, when elected, be an inhabitant of that State in which he shall be chosen.[1]

(3) Representatives and direct taxes (except income)[2] shall be apportioned among the several States which may be included within this Union according to their respective numbers, — *which shall be determined by adding to the whole number of free persons, including those bound to service for a term of years*—and excluding Indians not taxed, — **three fifths of all other persons.**[3] The actual enumeration shall be made within three years after the first meeting of the Congress of the United States, and within every subsequent term of ten years, in such manner as they shall by law direct. The number of Representatives shall not exceed one for every thirty thousand, but each State shall have at least one Representative. **[; and until such enumeration shall be made, the State of New Hampshire shall be entitled to choose 3; Massachusetts, 8; Rhode Island and Providence Plantations, 1; Connecticut, 5; New York, 6; New Jersey, 4; Pennsylvania, 8; Delaware, 1; Maryland, 6; Virginia, 10; North Carolina, 5; South Carolina, 5; and Georgia, 3.]**[4]

[1] See Amendment XIV, section 3.

[2] Insert. See Amendment XVI.

[3] See Amendment XIII and sections 1 and 2 of Amendment XIV.

[4] Obsolete since 1793.

(4) When vacancies happen in the representation from any State, the Executive Authority thereof shall issue writs of election to fill such vacancies.

(5) The House of Representatives shall choose their Speaker and other officers, and shall have the sole power of impeachment.

THE UNITED STATES SENATE

Sec. 3. (1) The Senate of the United States shall be composed of two Senators from each State, chosen by the — *Legislature* [5] — thereof, for six years; and each Senator shall have one vote.

[5] See Amendment XVII, paragraph 1.

(2) Immediately after they shall be assembled in consequence of the first election, they shall be divided as equally as may be into three classes. The seats of the Senators of the first class shall be vacated at the expiration of the second year, of the second class at the expiration of the fourth year, and of the third class at the expiration of the sixth year, so that one third may be chosen every second year: — *and if vacancies happen by resignation, or otherwise, during the recess of the legislature of any State, the Executive thereof may make temporary appointments until the next meeting of the legislature, which shall then fill such vacancies.* [6]

[6] See Amendment XVII, paragraph 2.

(3) No person shall be a Senator who shall not have attained to the age of thirty years, and been nine years a citizen of the United States, and who shall not, when elected, be an inhabitant of that State for which he shall be chosen. [7]

[7] See Amendment XIV, section 3.

(4) The Vice-President of the United States shall be President of the Senate, but shall have no vote unless they be equally divided.

(5) The Senate shall choose their other officers, and also a President pro tempore in the absence of the Vice President, or when he shall exercise the office of President of the United States.

(6) The Senate shall have the sole power to try all impeachments. When sitting for that purpose, they shall be on oath or affirmation. When the President of the United States is tried, the Chief Justice shall preside; and no person shall be convicted without the concurrence of two-thirds of the members present.

(7) Judgment in cases of impeachment shall not extend further than to removal from office, and disqualification to hold and enjoy any office of honor, trust, or profit under the United States; but the party convicted shall nevertheless be liable and subject to indictment, trial, judgment, and punishment, according to law.

ORGANIZATION OF CONGRESS

Sec. 4. (1) The times, places, and manner of holding elections for Senators and Representatives shall be prescribed in each State by the Legislature thereof; but the Congress may at any time by law make or alter such regulations, — *except as to the places of choosing Senators.*[8]

[8] See Amendment XVII.

(2) The Congress shall assemble at least once in every year, and such meetings shall be on the first Monday in December, unless they shall by law appoint a different day. **(See New Amendment, Annex XX.)**

Sec. 5. (1) Each house shall be the judge of the elections, returns, and qualifications of its own members, and a majority of each shall constitute a quorum to do business; but a smaller number may adjourn from day to-day, and may be authorized to compel the attendance of absent members in such manner and under such penalties as each house may provide.

(2) Each house may determine the rules of its proceedings, punish its members for disorderly behavior, and with the concurrence of two thirds expel a member.

(3) Each house shall keep a journal of its proceedings, and from time to time publish the same, excepting such parts as may in their judgment require secrecy; and the yeas and nays of the members of either house on any question shall, at the desire of one-fifth of those present, be entered on the journal.

(4) Neither house, during the session of Congress, shall, without the consent of the other, adjourn for more than three days, nor to any other place than that in which the two houses shall be sitting.

Sec. 6. (1) The Senators and Representatives shall receive a compensation for their services, to be ascertained by law, and paid out of the Treasury of the United States. They shall in all cases, except treason, felony, and breach of the peace, be privileged from arrest during their attendance at the session of their respective houses, and in going to and returning from the same; and for any speech or debate in either house they shall not be questioned in any other place.

(2) No Senator or Representative shall, during the time for which he was elected, be appointed to any civil office under the authority of the United States which shall have been created, or the emoluments whereof shall have been increased during such time; and no person holding any office under the United States shall be a member of either house during his continuance in office.

Sec. 7. (1) All bills for raising revenue shall originate in the House of Representatives, but the Senate may propose or concur with amendments, as on other bills.

(2) Every bill which shall have passed the House of Representatives and the Senate shall, before it becomes a law, be presented to the President of the United States; if he approve, he shall sign it, but if not, he shall return it, with his objections, to that house in which it shall have originated, who shall enter the objections at large on their journal, and proceed to reconsider it. If after such reconsideration two-thirds of that house shall agree to pass the bill, it shall be sent, together with the objections, to the other house, by which it shall likewise be reconsidered; and if approved by two-thirds of that house it shall become a law. But in all such cases the votes of both houses shall be determined by yeas and nays, and the names of the persons voting for and against the bill shall be entered on the journal of each house respectively. If any bill shall not be returned by the President, within ten days (Sundays excepted) after it shall have been presented to him, the same shall be a law in like manner as if he had signed it, unless the Congress by their adjournment prevent its return; in which case it shall not be a law.

(3) Every order, resolution, or vote to which the concurrence of the Senate and House of Representatives may be necessary (except on a question of adjournment) shall be presented to the President of the United States; and before the same shall take effect shall be approved by him, or being disapproved by him, shall be repassed by two-thirds of the Senate and House of Representatives, according to the rules and limitations prescribed in the case of a bill.

POWERS VESTED IN CONGRESS

Sec. 8. The Congress shall have power:
(1) To lay and collect taxes, duties, imposts, and excises, to pay the debts and provide for the common

defense and general welfare of the United States; but all duties, imposts, and excises shall be uniform throughout the United States.

(2) To borrow money on the credit of the United States.

(3) To regulate commerce with foreign nations, and among the several States, and with the Indian tribes.

(4) To establish a uniform rule of naturalization and uniform laws on the subject of bankruptcies throughout the United States.

(5) To coin money, regulate the value thereof, and of foreign coin, and fix the standard of weights and measures.

(6) To provide for the punishment of counterfeiting the securities and current coin of the United States.

(7) To establish post offices and post-roads.

(8) To promote the progress of science and useful arts by securing for limited times to authors and inventors the exclusive right to their respective writings and discoveries.

(9) To constitute tribunals inferior to the Supreme Court.

(10) To define and punish piracies and felonies committed on the high seas, and offenses against the law of nations.

(11) To declare war, grant letters of marque and reprisal, and make rules concerning captures on land and water.

(12) To raise and support armies, but no appropriation of money to that use shall be for a longer term than two years.

(13) To provide and maintain a navy.

(14) To makes rules for the government and regulation of the land and naval forces.

(15) To provide for calling forth the militia to execute the laws of the Union, suppress insurrection, and repel invasions.

(16) To provide for organizing, arming, and disciplining the militia, and for governing such part of them as may be employed in the service of the United States, reserving to the States respectively the appointment of the officers, and the authority of training the militia according to the discipline prescribed by Congress.

(17) To exercise exclusive legislation in all cases whatsoever over such district (not exceeding ten miles square) as may, by cession of particular States and the acceptance of Congress, become the seat of the Government of the United States, and to exercise like authority over all places purchased by the consent of the legislature of the State in which the same shall be, for the erection of forts, magazines, arsenals, dockyards, and other needful buildings; And

(18) To make all laws which shall be necessary and proper for carrying into execution the foregoing powers, and all other powers vested by this Constitution in the Government of the United States, or in any department or officer thereof.

RESTRAINTS FEDERAL AND STATE

Sec. 9. **[(1) The migration or importation of such persons as any of the States now existing shall think proper to admit shall not be prohibited by the Congress prior to the year one thousand eight hundred and eight, but a tax or duty may be imposed on such importation, not exceeding ten dollars for each person.]**[9]

[9] Obsolete since 1808.

(2) The privilege of the writ of habeas corpus shall not be suspended, unless when in cases of rebellion or invasion the public safety may require it.

(3) No bill of attainder or ex post facto law shall be passed.

(4) No capitation or other direct tax (except income)[10] shall be laid, unless in proportion to the census or enumeration hereinbefore directed to be taken.

[10] Insert. See Amendment XVI.

(5) No tax or duty shall be laid on articles exported from any State.

(6) No preference shall be given by any regulation of commerce or revenue to the ports of one State over those of another, nor shall vessels bound to or from one State be obliged to enter, clear, or pay duties in another.

(7) No money shall be drawn from the Treasury but in consequence of appropriations made by law; and a regular statement and account of the receipts and expenditures of all public money shall be published from time to time.

(8) No title of nobility shall be granted by the United States. And no person holding any office of profit or trust under them shall, without the consent of the Congress, accept of any present, emolument, office, or title of any kind whatever from any king, prince, or foreign state.

Sec. 10. (1) No State shall enter into any treaty, alliance, or confederation, grant letters of marque and reprisal, coin money, emit bills of credit, make anything but gold and silver coin a tender in payment of debts, pass any bill of attainder, ex post facto law, or law impairing the obligation of contracts, or grant any title of nobility.

(2) No State shall, without the consent of the Congress, lay any imposts or duties on imports or exports, except what may be absolutely necessary for executing its inspection laws, and the net produce of all duties and imposts, laid by any State on imports or exports, shall be for the use of the Treasury of the United States; and all

such laws shall be subject to the revision and control of the Congress.

(3) No State shall, without the consent of Congress, lay any duty of tonnage, keep troops or ships of war in time of peace, enter into any agreement or compact with another State, or with a foreign power, or engage in war, unless actually invaded or in such imminent danger as will not admit of delay.

ARTICLE II

THE EXECUTIVE DEPARTMENT

Section 1. (1) The executive power shall be vested in a President of the United States of America. He shall hold his office during the term of four years, and, together with the Vice-President, chosen for the same term, be elected as follows:

(2) Each State shall appoint, in such manner as the Legislature thereof may direct, a number of electors, equal to the whole number of Senators and Representatives to which the State may be entitled in the Congress; but no Senator or Representative, or person holding an office of trust or profit under the United States, shall be appointed an elector.[11]

[11] See Amendment XIV, Section 3.

(3) The electors shall meet in their respective States and vote by ballot for two persons, of whom one at least shall not be an inhabitant of the same State with themselves. And they shall make a list of all the persons voted for, and of the number of votes for each; which list they shall sign and certify, and transmit sealed to the seat of the government of the United States, directed to the President of the Senate. The President of the Senate shall, in the

presence of the Senate and House of Representatives, open all the certificates, and the votes shall then be counted. The person having the greatest number of votes shall be the President, if such number be a majority of the whole number of electors appointed; and if there be more than one who have such majority, and have an equal number of votes, then the House of Representatives shall immediately choose by ballot one of them for President; and if no person have a majority, then from the five highest on the list the said House shall, in like manner, choose the President.

But in choosing the President the votes shall be taken by States, the representation from each State having one vote; a quorum for this purpose shall consist of a member or members from two-thirds of the States, and a majority of all the States shall be necessary to a choice. In every case, after the choice of the President, the person having the greatest number of votes of the electors shall be the Vice-President. But if there should remain two or more who have equal votes, the Senate shall choose from them by ballot the Vice-President.[12]

[12] Supplanted by Amendment XII.

(4) The Congress may determine the time of choosing the electors, and the day on which they shall give their votes; which day shall be the same throughout the United States.

(5) No person except a natural born citizen **[or a citizen of the United States at the time of the adoption of this Constitution]** Obsolete shall be eligible to the office of President; neither shall any person be eligible to that office who shall not have attained to the age of thirty-five years and been fourteen years a resident within the United States.[15]

[15] See Amendment XIV, Section 3.

(6) In case of the removal of the President from office, or of his death, resignation, or inability to discharge the powers and duties of the said office, the same shall devolve on the Vice-President, and the Congress may by law provide for the case of removal, death, resignation, or inability, both of the President and Vice President, declaring what officer shall then act as President, and such officer shall act accordingly until the disability be removed or a President shall be elected.

(See New Amendment, Annex XXV.)

(7) The President shall, at stated times, receive for his services a compensation which shall neither be increased nor diminished during the period for which he shall have

been elected, and he shall not receive within that period any other emolument from the United States or any of them.

(8) Before he enter on the execution of his office, he shall take the following oath or affirmation:

"I do solemnly swear (or affirm) that I will faithfully execute the office of President of the United States, and will, to the best of my ability, preserve, protect, and defend the Constitution of the United States."

Sec. 2. (1) The President shall be commander-in-chief of the Army and Navy of the United States, and of the militia of the several States, when called into the actual service of the United States; he may require the opinion, in writing, of the principal officer in each of the executive departments, upon any subject relating to the duties of their respective offices; and he shall have power to grant reprieves and pardons for offenses against the United States, except in cases of impeachment.

(2) He shall have power, by and with the advice and consent of the Senate, to make treaties, provided two-thirds of the Senators present concur; and he shall

nominate, and by and with the advice and consent of the Senate shall appoint ambassadors, other public ministers and consuls, judges of the Supreme Court, and all other officers of the United States whose appointments are not herein otherwise provided for, and which shall be established by law; but the Congress may by law vest the appointment of such inferior officers as they think proper in the President alone, in the courts of law, or in the heads of departments.

(3) The President shall have power to fill up all vacancies that may happen during the recess of the Senate, by granting commissions which will expire at the end of their next session.

Sec. 3. He shall from time to time give to the Congress information of the state of the Union, and recommend to their consideration such measures as he shall judge necessary and expedient; he may, on extraordinary occasions, convene both houses, or either of them, and in case of disagreement between them, with respect to the time of adjournment, he may adjourn them to such time as he shall think proper; he shall receive ambassadors and other public ministers; he shall take care that the laws be faithfully executed, and shall commission all the officers of the United States.

Sec. 4. The President, Vice President, and all civil officers of the United States, shall be removed from office on impeachment for, and conviction of, treason, bribery, or other high crimes and misdemeanors.

ARTICLE III

THE JUDICIAL DEPARTMENT

Section 1. The judicial power of the United States shall be vested in one Supreme Court, and in such inferior courts as the Congress may from time to time ordain and establish. The judges, both of the Supreme and inferior courts, shall hold their offices during good behavior and shall, at stated times, receive for their services a compensation which shall not be diminished during their continuance in office.

Sec. 2. (1) The judicial power shall extend to all cases, in law and equity, arising under this Constitution, the laws of the United States, and treaties made, or which shall be made, under their authority; to all cases affecting ambassadors, other public ministers, and consuls; to all cases of admiralty and maritime jurisdiction; to controversies to which the United States shall be a party; to controversies between two or more States; — *between a State and citizens of another State;* [16] — between citizens of different States; between citizens of the same State claiming lands under grants of different States, and between a State, or the citizens thereof, and foreign states, — *citizens or subjects.* [16]

[16] See Amendment XI.

(2) In all cases affecting ambassadors, other public ministers, and consuls and those in which a State shall be party, the Supreme Court shall have original jurisdiction. In all the other cases before mentioned, the Supreme Court shall have appellate jurisdiction, both as to law and fact, with such exceptions and under such regulations as the Congress shall make.

(3) The trial of all crimes, except in cases of impeachment, shall be by jury; and such trial shall be held in the State where the said crimes shall have been committed; but when not committed within any State, the trial shall be at such place or places as the Congress may by law have directed.

Sec. 3. (1) Treason against the United States shall consist only in levying war against them, or in adhering to their enemies, giving them aid and comfort. No person shall be convicted of treason unless on the testimony of two witnesses to the same overt act, or on confession in open court.

(2) The Congress shall have power to declare the punishment of treason, but no attainder of treason shall work corruption of blood, or forfeiture, except during the life of the person attainted.

INTERSTATE AND FEDERAL RELATIONS

ARTICLE IV

RELATION OF THE STATES TO EACH OTHER

Section 1. Full faith and credit shall be given in each State to the public acts, records, and judicial proceedings of every other State. And the Congress may by general laws prescribe the manner in which such acts, records, and proceedings shall be proved, and the effect thereof.

Sec. 2. (1) The citizens of each State shall be entitled to all privileges and immunities of citizens in the several States.

(2) A person charged in any State with treason, felony, or other crime, who shall flee from justice, and be found in another State, shall, on demand of the Executive authority of the State from which he fled, be delivered up, to be removed to the State having jurisdiction of the crime.

(3) **[No person held to service or labor in one State, under the laws thereof, escaping into another shall, in consequence of any law or regulation therein, be discharged from such service or labor, but shall be delivered up on claim of the party to whom such service or labor may be due.]**[17]

[17] Obsolete. (See Amendment, Annex XIII.)

RELATION OF THE UNITED STATES TO STATES AND TERRITORIES

Sec. 3. (1) New States may be admitted by the Congress into this Union; but no new State shall be formed or erected within the jurisdiction of any other State, nor any State be formed by the junction of two or more States, or parts of States, without the consent of the Legislatures of the States concerned, as well as of the Congress.

(2) The Congress shall have power to dispose of and make all needful rules and regulations respecting the territory or other property belonging to the United States; and nothing in this Constitution shall be so construed as to prejudice any claims of the United States, or of any particular State.

Sec. 4. The United States shall guarantee to every State in this Union a republican form of government, and shall protect each of them against invasion, and, on application of the Legislature, or of the Executive (when the Legislature cannot be convened), against domestic violence.

GENERAL PROVISIONS

ARTICLE V

PROVISION FOR AMENDING THE CONSTITUTION

The Congress, whenever two-thirds of both houses shall deem it necessary, shall propose amendments this Constitution, or, on the application of the Legislatures of two-thirds of the several States, shall call a convention for proposing amendments, which in either case, shall be valid to all intents and purposes, as part of this Constitution, when ratified by the Legislatures of three-fourths of the several States, or by conventions in three-fourths thereof, as the one or the other mode of ratification may be proposed by the Congress; provided **[that no amendment which may be made prior to the year one thousand eight hundred and eight shall in any manner affect the first and fourth clauses in the Ninth Section of the First Article; and]** Obsolete that no State, without its consent, shall be deprived of its equal suffrage in the Senate.

ARTICLE VI

NATIONAL DEBTS

(1) All debts contracted and engagements entered into before the adoption of this Constitution shall be as valid against the United States under this Constitution as under the Confederation.

SUPREMACY OF THE NATIONAL GOVERNMENT

(2) This Constitution and the laws of the United States which shall be made in pursuance thereof and all treaties made, or which shall be made, under the authority of the United States, shall be the supreme law of the land; and the judges in every State shall be bound thereby, anything in the Constitution or laws of any State to the contrary notwithstanding.

PLEDGE — NO RELIGIOUS TEST

(3) The Senators and Representatives before mentioned, and the members of the several State Legislatures, and all executive and judicial officers, both of the United States and of the several States, shall be bound by oath or affirmation to support this Constitution; but no religious test shall ever be required as a qualification to any office or public trust under the United States.

ARTICLE VII

The ratification of the conventions of nine States shall be sufficient for the establishment of this Constitution between the States so ratifying the same.

Done in Convention by the unanimous consent of the States present the seventeenth day of September in the year of our Lord one thousand seven hundred and eighty-seven and of the independence of the United States of America the twelfth. In witness whereof we have hereunto subscribed our Names,

George Washington, President and delegate from Virginia

NEW HAMPSHIRE

John Langdon, Nicholas Gilman

MASSACHUSETTS

Nathaniel Gorham, Rufus King

CONNECTICUT

William Samuel Johnson, Roger Sherman

NEW YORK

Alexander Hamilton

NEW JERSEY

William Livingston, David Brearley, William Paterson
Jonathan Dayton

PENNSYLVANIA

Benjamin Franklin, Thomas Mifflin, Robert Morris
George Clymer, Thomas Fitzsimmons, Jared Ingersoll
James Wilson, Gouverneur Morris

DELAWARE

George Read, Gunning Bedford, John Dickinson
Richard Bassett, Jacob Broom

MARYLAND

James McHenry, Daniel of St. Thomas Jenifer
Daniel Carroll

VIRGINIA

John Blair, James Madison

NORTH CAROLINA

William Blount, Richard Dobbs Spaight
Hugh Williamson

SOUTH CAROLINA

John Rutledge, Charles Cotesworth Pinckney
Charles Pinckney, Pierce Butler

GEORGIA

William Few, Abraham Baldwin

Delegates Edmund Randolph and George Mason of Virginia and Elbridge Gerry of Massachusetts were present on the last day of the Convention but refused to sign the Constitution.

The following delegates were not present on the last day of the Convention, but a goodly portion of them were in favor of the Constitution: W. Oliver Ellsworth of Connecticut; William Churchill Houston of New Jersey; John Caleb Strong of Massachusetts; William Pierce and William Houston of Georgia; William Richardson Davie and Alexander Martin of North Carolina; James McClurg and George Wythe of Virginia; Robert Yates and W. John Lansing of New York; and John Francis Mercer and Luther Martin of Maryland.

Many people seem to have the impression that John Hancock, John Adams, Thomas Jefferson, and Patrick Henry were delegates to the Constitutional Convention, but none of them were.

AMENDMENTS

The first ten, proposed September 25, 1789; adopted June 15, 1790

ARTICLE I

Congress shall make no law respecting an establishment of religion, or prohibiting the free exercise thereof; or abridging the freedom of speech, or of the press; or the right of the people peaceably to assemble, and to petition the government for a redress of grievances.

ARTICLE II

A well-regulated militia being necessary to the security of a free State, the right of the people to keep and bear arms shall not be infringed.

ARTICLE III

No soldier shall, in time of peace, be quartered in any house without the consent of the owner; nor in time of war but in a manner to be prescribed by law.

ARTICLE IV

The right of the people to be secure in their persons, houses, papers and effects, against unreasonable searches and seizures, shall not be violated, and no warrants shall issue but upon probable cause, supported by oath or affirmation, and particularly describing the place to be searched, and the persons or things to be seized.

ARTICLE V

No person shall be held to answer for a capital or otherwise infamous crime, unless on a presentment or indictment of a grand jury, except in cases arising in the land or naval forces, or in the militia, when in actual service in time of war or public danger; nor shall any person be subject for the same offense to be twice put in jeopardy of life or limb; nor shall be compelled in any criminal case to be a witness against himself, nor be deprived of life, liberty, or property, without due process of law; nor shall private property be taken for public use, without just compensation.

ARTICLE VI

In all criminal prosecutions the accused shall enjoy the right to a speedy and public trial, by an impartial jury of the State and district wherein the crime shall have been committed, which district shall have been previously ascertained by law, and to be informed of the nature and cause of the accusation; to be confronted with the witnesses against him; to have compulsory process for obtaining witnesses in his favor, and to have the assistance of counsel for his defense.

ARTICLE VII

In suits at common law, where the value in controversy shall exceed twenty dollars, the right of trial by jury shall be preserved, and no fact tried by a jury shall be otherwise re-examined in any court of the United States than according to the rules of the common law.

ARTICLE VIII

Excessive bail shall not be required, nor excessive fines imposed, nor cruel and unusual punishments inflicted.

ARTICLE IX

The enumeration in the Constitution of certain rights shall not be construed to deny or disparage others retained by the people.

ARTICLE X

The powers not delegated to the United States by the Constitution, nor prohibited by it to the States, are reserved to the States respectively, or to the people.

ARTICLE XI

Proposed September 6, 1794; adopted January 8, 1798

The judicial power of the United States shall not be construed to extend to any suit in law or equity, commenced or prosecuted against one of the United States by citizens of another State, or by citizens or subjects of any foreign state.

(This amendment modifies paragraph 1, section 2, of Article III.)

ARTICLE XII

Proposed December 12, 1803; adopted September 25, 1804

The electors shall meet in their respective States, and vote by ballot for President and Vice-President, one of whom at least shall not be an inhabitant of the same State with themselves; they shall name in their ballots the person voted for as President, and in distinct ballots the person voted for as Vice-President; and they shall make distinct lists of all persons voted for as President, and of all persons voted for as Vice-President, and of the number of votes for each, which list they shall sign and certify, and transmit, sealed, to the seat of the Government of the United States, directed to the President of the Senate; the President of the Senate shall, in the presence of the Senate and House of Representatives, open all the certificates, and the votes shall then be counted; the person having the greatest number of votes for President shall be the President, if such number be a majority of the whole number of electors appointed; and if no person have such majority, then from the persons having the highest numbers, not exceeding three, on the list of those voted for as President, the House of Representatives shall choose immediately, by ballot, the President. But in choosing the President, the votes shall be taken by States, the representation from each State having one vote; a quorum for this purpose shall consist of a member or members from two-thirds of the States, and a majority of all the States shall be necessary to a choice. *And if the House of Representatives shall not choose a President, whenever the right of choice shall devolve upon them, before the fourth day of March next following, then the Vice-President shall act as President, as in the case of the death or other constitutional disability of the President.*[18] The person having the greatest number of votes as Vice-President shall be the Vice-President if such number be a majority of the whole number of electors appointed, and if no person have a majority, then from the two highest numbers on the list the Senate shall choose the Vice-President; a quorum for the purpose shall consist of two-thirds of the whole number of Senators, and a majority of the whole number shall be necessary to a choice. But no person constitutionally ineligible to the office of President shall be eligible to that of Vice-President of the United States.

[18] (See New Amendment, Annex XX.)

(This amendment supplants paragraph 3, section 1, of Article II.)

ARTICLE XIII

Proposed February 1, 1865; adopted December 18, 1865

Section 1. Neither slavery nor involuntary servitude, except as a punishment for crime, whereof the party shall have been duly convicted, shall exist within the United States, or any place subject to their jurisdiction.

Sec. 2. Congress shall have power to enforce this article by appropriate legislation.

ARTICLE XIV

Proposed June 16, 1866; adopted July 21, 1868

Section 1. All persons born or naturalized in the United States, and subject to the jurisdiction thereof, are citizens of the United States and of the State wherein they reside. No State shall make or enforce any law which shall abridge the privileges or immunities of citizens of the United States; nor shall any State deprive any person of life, liberty, or property, without due process of law; nor deny to any person within its jurisdiction the equal protection of the laws.

Sec. 2. Representatives shall be apportioned among the several States according to their respective numbers, counting the whole number of persons in each State, excluding Indians not taxed. But when the right to vote at any election for the choice of electors for President and Vice President of the United States, Representatives in Congress, the executive and judicial officers of a State, or the members of the legislature thereof, is denied to any of the *male* [20] inhabitants of such State, *being twenty-one years of age* [21] and citizens of the United States, or in any way abridged, except for participation in rebellion, or other crime, the basis of representation therein shall be reduced in the proportion which the number of such male citizen shall bear to the whole number of male citizens twenty-one years of age in such State.

[20] (See New Amendment, Annex XIX.)

[21] (See New Amendment, Annex XXVI.)

(Sections 1 and 2 of this amendment modify paragraph 3, section 2, of Article I.)

Sec. 3. No person shall be a Senator, or Representative in Congress, or elector of President and Vice-President, or hold any office, civil or military, under the United States, or under any State, who, having previously taken an oath as a member of Congress, or as an officer of the United States, or as a member of any State legislature, or as an executive or judicial officer of any State, to support the

Constitution of the United States, shall have engaged in insurrection or rebellion against the same, or given aid or comfort to the enemies thereof. But Congress may, by a vote of two-thirds of each house, remove such disability.

(Section 3 of this amendment supplements paragraph 2, section 2, of Article I; paragraph 3, section 3, of Article I; paragraph 2, section 1, of Article II; and, paragraph 5, section 1, of Article II.)

Sec. 4. The validity of the public debt of the United States, authorized by law, including debts incurred for payment of pensions and bounties for services in suppressing insurrection or rebellion, shall not be questioned. But neither the United States nor any State shall assume or pay any debt or obligation incurred in aid of insurrection or rebellion against the United States, or any claim for the loss or emancipation of any slave; but all such debts, obligations and claims shall be held illegal and void.

Sec. 5. The Congress shall have power to enforce, by appropriate legislation, the provisions of this article.

ARTICLE XV

Proposed February 27, 1869; adopted March 30, 1870

Section 1. The right of citizens of the United States to vote shall not be denied or abridged by the United States, or by any State, on account of race, color, or previous condition of servitude.

Sec. 2. The Congress shall have power to enforce this article by appropriate legislation.

(This amendment supplements paragraph 1, section 2, of Article I.)

245

ARTICLE XVI

Proposed July 31, 1909;
adopted February 25, 1913

The Congress shall have power to lay and collect taxes on incomes, from whatever source derived, without apportionment among the several States, and without regard to any census or enumeration.

(This amendment modifies paragraph 3, section 2, of Article I, and paragraph 4, section 9, of Article I.)

ARTICLE XVII

Proposed May 15, 1912;
adopted May 81, 1913

(1) The Senate of the United States shall be composed of two Senators from each State, elected by the people thereof for six years, and each Senator shall have one vote. The electors in each State shall have the qualifications requisite for electors of the most numerous branch of the State legislatures.

(Paragraph 1 of this amendment modifies paragraph 1, section 3, of Article I, and paragraph 1, section 4, of Article I.)

(2) When vacancies happen in the representation of any State in the Senate, the executive authority of such State shall issue writs of election to fill such vacancies: *Provided,* That the legislature of any State may empower the executive thereof to make temporary appointments until the people fill the vacancies by election, as the legislature may direct.

(Paragraph 2 of this amendment modifies paragraph 2, section 3, of Article I.)

(3) This amendment shall not be so construed as to affect the election or term of any Senator chosen before it becomes valid as part of the Constitution.

ARTICLE XVIII

Proposed December 19, 1917; adopted January 29, 1919

Section 1. After one year from the ratification of this article the manufacture, sale, or transportation of intoxicating liquors within, the importation thereof into, or the exportation thereof from the United States and all territory subject to the jurisdiction thereof for beverage purposes is hereby prohibited.

Sec. 2. The Congress and the several States shall have concurrent power to enforce this article by appropriate legislation.

Sec. 3. This article shall be inoperative unless it shall have been ratified as an amendment to the Constitution by the legislatures of the several States, as provided in the Constitution, within seven years from the date of submission hereof to the States by the Congress.

ARTICLE XIX

Proposed June 6, 1919; adopted August 26, 1920

Section 1. The right of citizens of the United States to vote shall not be denied or abridged by the United States or by any State on account of sex.

Sec. 2. Congress shall have power to enforce this article by appropriate legislation.

(This amendment supplements paragraph 1, section 2, of Article I.)

The following articles of amendments were added after the publishing of TM 2000-25.

ARTICLE XX

Proposed March 2, 1932; adopted January 20, 1933

Section 1. The terms of the President and Vice President shall end at noon on the 20th day of January, and the terms of Senators and Representatives at noon on the 3d day of January, of the years in which such terms would have ended if this article had not been ratified; and the terms of their successors shall then begin.

Sec. 2. The Congress shall assemble at least once in every year, and such meeting shall begin at noon on the 3d day of January, unless they shall by law appoint a different day.

Sec. 3. *If, at the time fixed for the beginning of the term of the President, the President elect shall have died, the Vice President elect shall become President. If a President shall not have been chosen before the time fixed for the beginning of his term, or if the President elect shall have failed to qualify, then the Vice President elect shall act as President until a President shall have qualified; and the Congress may by law provide for the case wherein neither a President elect nor a Vice President elect shall have qualified, declaring who shall then act as President, or the manner in which one who is to act shall be selected, and such person shall act accordingly until a President or Vice President shall have qualified.*[22]

[22] (See New Amendment XXV.)

Sec. 4. The Congress may by law provide for the case of the death of any of the persons from whom the House of

Representatives may choose a President whenever the right of choice shall have devolved upon them, and for the case of the death of any of the persons from whom the Senate may choose a Vice President whenever the right of choice shall have devolved upon them.

Sec. 5. Sections 1 and 2 shall take effect on the 15th day of October following the ratification of this article.

Sec. 6. This article shall be inoperative unless it shall have been ratified as an amendment to the Constitution by the legislatures of three-fourths of the several states within seven years from the date of its submission.

ARTICLE XXI

Proposed February 20, 1933; adopted December 5, 1933

Section 1. The eighteenth article of amendment to the Constitution of the United States is hereby repealed.

Sec. 2. The transportation or importation into any state, territory, or possession of the United States for delivery or use therein of intoxicating liquors, in violation of the laws thereof, is hereby prohibited.

Sec. 3. This article shall be inoperative unless it shall have been ratified as an amendment to the Constitution by conventions in the several states, as provided in the Constitution, within seven years from the date of the submission hereof to the states by the Congress.

ARTICLE XXII

Proposed February 6, 1947;
adopted March 21, 1947

Section 1. No person shall be elected to the office of the President more than twice, and no person who has held the office of President, or acted as President, for more than two years of a term to which some other person was elected President shall be elected to the office of the President more than once. But this article shall not apply to any person holding the office of President when this article was proposed by the Congress, and shall not prevent any person who may be holding the office of President, or acting as President, during the term within which this article becomes operative from holding the office of President or acting as President during the remainder of such term.

Sec. 2. This article shall be inoperative unless it shall have been ratified as an amendment to the Constitution by the legislatures of three-fourths of the several states within seven years from the date of its submission to the states by the Congress.

ARTICLE XXIII

Proposed S.Res.39 - 86th Congress, 1959;
adopted June 16, 1960

Section 1. The District constituting the seat of government of the United States shall appoint in such manner as the Congress may direct:

A number of electors of President and Vice President equal to the whole number of Senators and

Representatives in Congress to which the District would be entitled if it were a state, but in no event more than the least populous state; they shall be in addition to those appointed by the states, but they shall be considered, for the purposes of the election of President and Vice President, to be electors appointed by a state; and they shall meet in the District and perform such duties as provided by the twelfth article of amendment.

Sec. 2. The Congress shall have power to enforce this article by appropriate legislation.

ARTICLE XXIV

Proposed pre-1960; adopted August 27, 1962

Section 1. The right of citizens of the United States to vote in any primary or other election for President or Vice President, for electors for President or Vice President, or for Senator or Representative in Congress, shall not be denied or abridged by the United States or any state by reason of failure to pay any poll tax or other tax.

Sec. 2. The Congress shall have power to enforce this article by appropriate legislation.

ARTICLE XXV

Proposed January 6, 1965; adopted July 6, 1965

Section 1. In case of the removal of the President from office or of his death or resignation, the Vice President shall become President.

Sec. 2. Whenever there is a vacancy in the office of the Vice President, the President shall nominate a Vice

251

President who shall take office upon confirmation by a majority vote of both Houses of Congress.

Sec. 3. Whenever the President transmits to the President pro tempore of the Senate and the Speaker of the House of Representatives his written declaration that he is unable to discharge the powers and duties of his office, and until he transmits to them a written declaration to the contrary, such powers and duties shall be discharged by the Vice President as Acting President.

Sec. 4. Whenever the Vice President and a majority of either the principal officers of the executive departments or of such other body as Congress may by law provide, transmit to the President pro tempore of the Senate and the Speaker of the House of Representatives their written declaration that the President is unable to discharge the powers and duties of his office, the Vice President shall immediately assume the powers and duties of the office as Acting President.

Thereafter, when the President transmits to the President pro tempore of the Senate and the Speaker of the House of Representatives his written declaration that no inability exists, he shall resume the powers and duties of his office unless the Vice President and a majority of either the principal officers of the executive department or of such other body as Congress may by law provide, transmit within four days to the President pro tempore of the Senate and the Speaker of the House of Representatives their written declaration that the President is unable to discharge the powers and duties of his office. Thereupon Congress shall decide the issue, assembling within forty-eight hours for that purpose if not in session. If the Congress, within twenty-one days after receipt of the latter written declaration, or, if Congress is not in session, within twenty-one days after Congress is required to assemble, determines by two-thirds vote of both Houses that the President is unable to discharge the powers and duties of his office, the Vice President shall continue to discharge the same as Acting President; otherwise, the President shall resume the powers and duties of his office.

ARTICLE XXVI

Proposed 1968;
adopted March 23, 1971

Section 1. The right of citizens of the United States, who are 18 years of age or older, to vote, shall not be denied or abridged by the United States or any state on account of age.

Sec. 2. The Congress shall have the power to enforce this article by appropriate legislation.

ARTICLE XXVII

Proposed/adopted June 8, 1789;
ratified May 7, 1992

No law, varying the compensation for the services of the Senators and Representatives, shall take effect, until an election of Representatives shall have intervened.

On October 22, 1787, a little over a month after the Constitution was adopted by the Convention at Philadelphia, the hard-headed and many-sided Benjamin Franklin, when he was nearly 82 years of age, wrote a suggestion to a friend in Europe, which is still worthy of consideration, as follows:

"I send you enclosed the proposed new Federal Constitution for these States. I was engaged four months of the last summer in the Convention that formed it. It is now sent by Congress to the several States for their confirmation. If it succeeds, I do not see why you might not in Europe form a Federal Union and one grand republic of all its different states and kingdoms; by means of a like convention; for we had many interests to reconcile." — **Benjamin Franklin**

"Whatever may be the judgment pronounced on the competency of the architects of the Constitution, or whatever may be the destiny of the edifice prepared by them, I feel it a duty to express my profound and solemn conviction, derived from my intimate opportunity of observing and appreciating the views of the Convention collectively and individually, that there never was an assembly of men, charged with a great and arduous trust, who were more pure in their motives, or more exclusively or anxiously devoted to the object committed to them." — **James Madison.**

"Hold on, my friends, to the Constitution of your country, and the government established under it. Perform those duties which are present, plain and positive. Respect the laws of your country, uphold our American institutions as far as you are able, consult the chart and the compass: as if our united constitutional American liberty were in some degree committed to your charge, keep her, so far as it depends on you, clear of the breakers"—**Daniel Webster.**

Francois Guizot, the French philosopher, historian, and prime minister, once asked James Russell Lowell, noted author and poet, "How long do you think the American Republic will endure?" Lowell replied, "So long as the ideas of its founders continue to be dominant."

BIBLIOGRAPHY

"The Federalists"--------------------- Hamilton, Madison, Jay.

"The Constitution Explained"---------------- Harry Atwood.

"Formation of the Union of the American States"-----------
---------------------------------- Superintendent of Documents,
Government Printing Office, Washington, D. C.

"The Constitution of the United States,
    Its Sources and Application"---------- Thomas J. Norton.

"History of the United States"---------------------- McMaster.

INDEX

Subjects as published in 1928,
adapted to page numbers in this book.

A

B

C

C (continued)

C (continued)

D

E

F

G

H

I

J

K

L

M

N

O

P

Q

R

S

S (continued)

T

V

W

James L. Tippins

[A.G. 014.33 (4-28-28).]

BY ORDER OF THE SECRETARY OF WAR:

C. P. SUMMERALL,
Major General,
Chief of Staff.

OFFICIAL:
LUTZ WAHL,
Major General,
The Adjutant General.

ADDITIONAL COPIES
OF THIS PUBLICATION MAY BE PROCURED FROM
THE SUPERINTENDENT OF DOCUMENTS
U.S. GOVERNMENT PRINTING OFFICE
WASHINGTON, D. C.
AT
30 CENTS PER COPY

ABOUT THE COVER

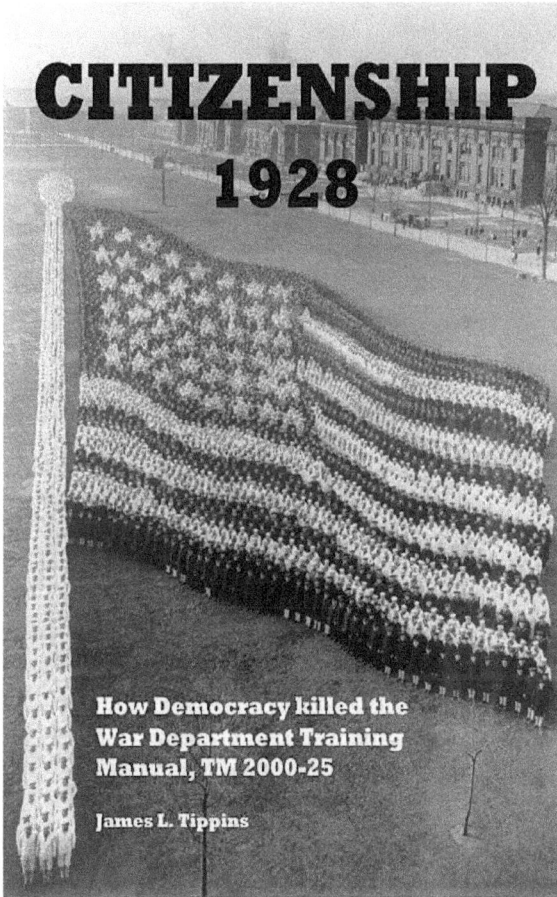

Cover © 2021 by James L. Tippins

How many sailors do you think are in this picture? A hundred? A thousand? Five thousand? How about TEN thousand?

Kristi Finefield provides a great explanation. Kristi is a Reference Specialist at the Prints & Photographs Division of the Library of Congress. Much thanks, Kristi!

Formation photograph of the American flag. Photo copyrighted 1917.
In the Public Domain.
www.loc.gov/item/2003655430/

Kristi: When I take a photo of a group of people, the challenges are familiar to most of us. Are the shorter people in the front so they can be seen? Is everyone's face visible? Are they smiling? Is everyone looking at the camera? And inevitably, someone still has their eyes closed in the final product.

Kristi: Now, multiply the number of people you're wrangling by *thousands* and add a new set of challenges, including color-coordinated outfits and the mathematics of perspective on a grand scale. You now have some idea what it took for Arthur Mole and John Thomas to capture the American flag image above.

Kristi: Taken in 1917 at the Naval Training Station in Great Lakes, Illinois, this photo includes about 10,000 U.S. Navy sailors dressed in either white or navy-blue uniforms to create what Arthur Mole called a "living photograph." His partner, John Thomas, stayed on the ground, coordinating the thousands of soldiers' placement, while Mole took his place on top of a 75-foot tall tower with a camera in hand.

Kristi: Mole & Thomas carefully planned out this and their future formation photographs beforehand, deciding how many troops to place in each location to counteract the perspective effects. The illusion is very effective, and it takes some very close looking to detect the tricks of the photo. For example, the entire flagpole contains 560 men, while the small ball on top, the furthest from the camera, required nearly 300 men alone. And the entire flag is a mere 73 feet wide at the bottom, but nearly 300 feet at the top, with scores more men standing along the rippling top stripe, so the flag appears to be waving in the breeze. (Narriative Credit Kristi Finefield of the Prints & Photographs Division, LOC.)

Here is another example. Maybe you can guess the number of soldiers because it isn't in the photo data!

James L. Tippins

SOMETHING VERY SMALL TO CONSIDER

During the final years of my USAF career, I was fortunate to witness every Space Shuttle launch at Kennedy Space Center. I had credentials that granted me access almost everywhere, including Launch Complex 39B.

The launch of the Hubble Space Telescope (HST) in April 1990 was the most significant non-classified science mission ever attempted, and I wasn't part of the launch team. What should I do?

I spent my extra time in the Payload Changeout Room at the pad, observing work progress on pre-launch closeouts. I gathered potential FOD materials and helped the team wherever I could, but stayed out of the way. I'm not listed as a launch team member, but I feel I contributed in a small way.

HST would become one of NASA's most significant missions ever launched.

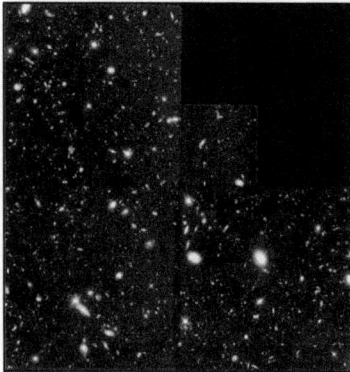

Robert Williams directed the telescope to devote ten consecutive days to imaging an "empty" place in the sky. To understand this area's size, stretch your arm out and pretend to hold a grain of sand between your fingers. It was an area that small.

When the resulting picture analysis was completed, about **3,000 galaxies** appeared in the photo. Imagine the rest of the Universe! We may be the only life that exists, and we are fortunate to live in the greatest nation on the planet.

ABOUT THE COPYEDITOR

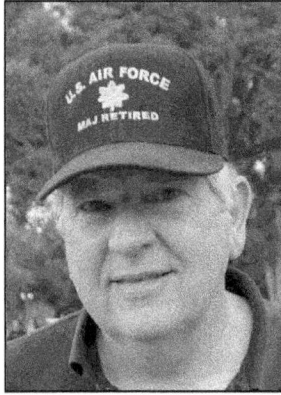

James L. Tippins
Rockledge, Florida, USA

Jim Tippins was born in Florida and grew up in Daytona Beach during the vibrant days of the 50s and 60s. He had a typical childhood in a traditional middle-class American family. His parents were firm, loving, and genuinely American.

As a teenager, he joined the Civil Air Patrol and then the US Air Force at age nineteen. He served five years as an enlisted member and another fifteen as an officer, retiring as a Major in 1991. The rest of his career involved Information Technology, a skill he often uses.

After discovering TM 2000-25 online, he sought to bring it back into the public eye. Every American should read about our history through the viewpoint of the constitutional historian Harry Atwood. International lovers of freedom could also gain valuable insights from here. Thank you for reading this book.

Notes

Notes

Notes